Animal Welfare

The Universities Federation for Animal Welfare

UFAW, founded in 1926, is an internationally recognised, independent, scientific and educational animal welfare charity that promotes high standards of welfare for farm, companion, laboratory and captive wild animals, and for those animals with which we interact in the wild. It works to improve animals' lives by:

- Funding and publishing developments in the science and technology that underpin advances in animal welfare;
- Promoting education in animal care and welfare;
- Providing information, organising meetings and publishing books, videos, articles, technical reports and the journal Animal Welfare;
- Providing expert advice to government departments and other bodies and helping to draft and amend laws and guidelines;
- Enlisting the energies of animal keepers, scientists, veterinarians, lawyers and others who care about animals.

Improvements in the care of animals are not now likely to come of their own accord, merely by wishing them: there must be research...and it is in sponsoring research of this kind, and making its results widely known, that UFAW performs one of its most valuable services.

Sir Peter Medawar CBE FRS, 8 May 1957

Nobel Laureate (1960), Chairman of the UFAW Scientific Advisory Committee (1951–1962)

UFAW relies on the generosity of the public through legacies and donations to carry out its work, improving the welfare of animals now and in the future. For further information about UFAW and how you can help promote and support its work, please contact us at the following address:

Universities Federation for Animal Welfare
The Old School, Brewhouse Hill, Wheathampstead, Herts AL4 8AN, UK
Tel: 01582 831818 Website: www.ufaw.org.uk
Email: ufaw@ufaw.org.uk

UFAW's aim regarding the UFAW/Wiley-Blackwell Animal Welfare book series is to promote interest and debate in the subject and to disseminate information relevant to improving the welfare of kept animals and of those harmed in the wild through human agency. The books in this series are the works of their authors, and the views they express do not necessarily reflect the views of UFAW.

Animal Welfare

Understanding Sentient Minds and Why it Matters

John Webster, MA, Vet MB, PhD, DVM (Hon)
Emeritus Professor of Animal Husbandry
University of Bristol School of Veterinary Science
Bristol Centre for Animal Behaviour
and Welfare Science
Bristol, UK

WILEY Blackwell

This edition first published 2022

Series Editors: Robert C. Hubrecht and Huw Golledge.

Registered Offices
John Wiley & Sons, Inc., 111 River Street, Hoboken, NJ 07030, USA
John Wiley & Sons Ltd, The Atrium, Southern Gate, Chichester, West Sussex, PO19 8SQ, UK

Editorial Office
9600 Garsington Road, Oxford, OX4 2DQ, UK

For details of our global editorial offices, customer services, and more information about Wiley products visit us at www.wiley.com.

Wiley also publishes its books in a variety of electronic formats and by print-on-demand. Some content that appears in standard print versions of this book may not be available in other formats.

Library of Congress Cataloging-in-Publication Data

Names: Webster, John, 1938– author.
Title: Animal welfare. Understanding sentient minds and why it matters / John Webster.
Other titles: Understanding sentient minds and why it matters | UFAW animal welfare series
Description: Hoboken, NJ : Wiley-Blackwell, 2022. | Series: UFAW/Wiley-Blackwell animal welfare book series
Identifiers: LCCN 2022001000 (print) | LCCN 2022001001 (ebook) | ISBN 9781119857068 (paperback) | ISBN 9781119857075 (adobe pdf) | ISBN 9781119857082 (epub)
Subjects: MESH: Animal Welfare
Classification: LCC HV4708 (print) | LCC HV4708 (ebook) | NLM HV 4708 | DDC 636.08/32–dc23/eng/20220204
LC record available at https://lccn.loc.gov/2022001000
LC ebook record available at https://lccn.loc.gov/2022001001

Cover Design: Wiley
Cover Images: © Judy Tomlinson/Getty Images; Agus Mahmuda/Getty Images; Paul Souders/ Getty Images; Pakkawit Anantaya/Getty Images

Set in 9.5/12pt Sabon by Straive, Pondicherry, India

Printed and bound by CPI Group (UK) Ltd, Croydon, CR0 4YY

C9781119857068_290322

To a startled mouse, and a serene rat, who gave me the idea.

Contents

About the Author

John Webster

John Webster is Emeritus Professor of Animal Husbandry, University of Bristol School of Veterinary Science and founder of the internationally recognised Bristol Centre for Animal Behaviour and Welfare Science. His previous books include Animal Welfare: A Cool Eye towards Eden, Limping towards Eden and Animal Husbandry Regained.

Preface

This is the third Animal Welfare book that I have written in a series published by the Universities Federation for Animal Welfare (UFAW). Each book has been designed to reflect stages on a journey. The first in the series, 'A Cool Eye towards Eden' (1994) imagined an ideal world for all sentient animals defined by a perfect expression of the five freedoms, acknowledged that while this was unachievable, it was the right way to travel and outlined the path. It posed two questions: 'How is it for them?' and 'What can we do for them?' It addressed the physiological and psychological (body and mind) determinants of animal welfare and outlined an approach to promoting their wellbeing on the farm, in the laboratory and in the home. It closed with a restatement of the view of Eden in the words of Albert Schweitzer 'Until he extends his circle of compassion to all living things, man himself will not find peace.'

My second book, 'Limping towards Eden' (2005), was written at a time of intensive activity in animal welfare science and sought to incorporate this new knowledge into good practice in regard to our approach to, and treatment of, animals in our care. Its closing words were more downbeat. We are on an endless journey towards an impossible dream but 'the path of duty lies in what is near. We may never expect to see our final destination but, for those who are prepared to open their eyes, the immediate horizon is full of promise.'

What these first two books had in common is that both were primarily concerned with practical solutions to problems of animal welfare based on better understanding of their physical and behavioural needs. While this latest book is based on the same principles of respect for animals- my views haven't changed – the subject matter is very different. It is a very different journey: a voyage of exploration into the animal mind, the nature and extent of sentience and consciousness in different species and how these have been shaped by the challenges of their environment. In Part 1, I set out the operation manual and toolbox of special senses, physical and mental faculties available to all sentient animals as their instinctive birth-right and explore how animals with the properties of a sentient mind are able to build on this birth-right and develop survival and social strategies to promote their wellbeing and perfect their use of the tools available to them at birth. In Part 2 I review how sentient minds have been shaped through adaptation to their natural environments. I consider first animals of the waters and the air; least subject to interference from that most invasive of terrestrial species, mankind. Terrestrial mammals are considered in two groups, first the herbivores, carnivores and omnivores of the open plains, then the animals of the forests and jungles, with special attention to the physical and social skills needed for life in the three dimensions of the tree canopy. In this section there is some repetition of the themes introduced in Part 1. This is necessary to put these properties of sentience into their environmental context and (under pressure) I have tried to keep them as brief as possible. The last chapter in this section considers how the finely tuned balance between inherited and acquired senses and skills in different natural environments has been affected by domestication, giving special concern to those animals whose behaviour has been most affected by human interference, dogs, pigs and horses. Since my aim is to look at animals through their eyes, not ours, human attitudes are kept, wherever possible, off the page. However, in the final chapter, 'Our duty of care', I address the second clause in my title. I examine human attitudes and actions in regard to other animals on the basis that animals with sentient minds have feelings that matter to them, so they should matter to us too. I review how we can apply our understanding of the sentient mind to meet our responsibilities and govern our approach to animals in the home, on farms, in sport and entertainment, in laboratories and in the wild. This brief voyage into the animal mind is based, wherever possible on evidence from science and sound practice. However, it regularly ventures into uncharted waters so contains almost as much speculation as hard evidence and makes no claim to be definitive. It has not been my aim to present a comprehensive, balanced review of existing knowledge and understanding in relation to animal minds but to stimulate the desire in your mind to learn more. The better we understand our fellow mortals, the more likely it is that we can be good neighbours.

Acknowledgements and Apologies

The aim of all educational books is to contribute to knowledge and understanding. My book is addressed to all who care for sentient animals, which is a much broader reading public than just academics and students diligently studying animal welfare as a part of their formal education. However, it must carry the authority that comes from diligent research. Just as the animals acquire different skills to meet different challenges, being expert in some things, ignorant of others, so too the academics. With this I mind, I must acknowledge at the outset that this book is crazily ambitious in scope. It carries the scent of *'life, the universe and everything'* in that it seeks to embrace the full extent of our knowledge and understanding of the sentient minds of animals in the context of the full range of challenges and opportunities presented by life on earth. Moreover, I acknowledge, it abounds in speculation. Many millions of words have been written by scientists, philosophers and fellow travellers seeking to understand the minds of animals, how they are shaped by their environments and how these are linked to the workings of the brain. My reading of this is wide but, inevitably less wide than it could be. If I were to attempt to acknowledge the sources of every assertion made in this book, the list of references would be longer than the book itself and even then, I would be guilty of omitting at least as many seminal references as I included. Moreover, a comprehensive list of references intended to direct a library search no longer carries the importance it once had. In recent years, my research, like that of everybody else, has been made so much quicker and more comprehensive by the reading and careful interpretation of on-line information from sources such as Google Scholar and Wikipedia. Readers wishing to confirm or contest my assertions in regard to well-documented issues, or simply seek further and better particulars, should be able to get access to almost all my sources in two to three clicks. In the section 'Further Reading', I list a number of good books that expand on some of the big topics presented here in brief. Most of the specific references listed under further reading deal with material taken from a specific scientific communication. When I speculate beyond the constraints of the literature and cannot therefore stand on the shoulders of others, I strive always to conform to first principles of science that apply across a broad spectrum so do not need the support of written evidence relating to every possible circumstance. Water runs downhill, wherever one happens to be.

I have spent over 60 years working with animals, thinking about animals, discussing animals with wise colleagues, writing and teaching about animals. I cannot possibly acknowledge by name all those who have guided and developed my thoughts: distinguished colleagues who have enriched my understanding; razor-sharp students who have challenged my convictions. I have therefore taken the easy option and never (well, almost never) named names. Those of you who read this book and recognise that I am talking about you, please accept my heartfelt thanks. I would make one exception to my policy of not naming names. I am deeply indebted to Birte Nielsen of UFAW, who has conscientiously and wisely helped to knock this manuscript into shape, purged me of repetitions and reined me in whenever my imagination was getting out of hand.

Part 1

The Sentient Mind: Skills and Strategies

I think I could turn and live with animals...I look at them long and long.
They do not sweat and whine about their condition.
They do not lie awake at night and whine about their sins.
Not one is dissatisfied. Not one is demented by the mania of owning things.
Not one is respectable or unhappy over the whole earth.

From Walt Whitman, Song of Myself

Setting the Scene

1

Some years ago, I took part in a late night, 'bear-pit' style television debate on the rights and wrongs of fishing. My role was to present scientific evidence as to whether fish can experience pain and fear. In brief, the evidence shows they can. After I had outlined the results of this work, a member of the audience got up and said 'This is all rubbish. These scientists don't know what they are talking about. I have been fishing all my life and I know for certain that fish don't feel anything'. He then added 'What sort of fish were they anyway? and when I said 'carp' he said: 'Ah well, carp are clever buggers'. These four words encapsulate the need for this book. We sort of assume animals have minds. We may even think we understand the meaning of sentience but most of us don't give it much thought, because, for most of us, most animals don't much matter.

This book is written for those for whom it matters a lot. My central aim is to equip you to seek a better understanding of the minds of sentient animals. To this end, it will not only give an outline review of existing knowledge relating to the mental processes that determine animal behaviour and welfare but also offer suggestions and guidance on how to approach subjects where we know little or have been relying on easy preconceptions. Those of us who embark on the scientific study of animal welfare, their needs, their behaviour and their motivation, are cautioned to avoid the fallacy of anthropomorphism: the fallacy of ascribing human characteristics to other animals. However,

Animal Welfare: Understanding Sentient Minds and Why it Matters, First Edition. John Webster.

I suggest at the outset, that it is valid to apply a principle of reverse anthropomorphism that asks not 'how would this chicken, cow, horse' feel if it were me but how would I feel if I were one of them?' As we shall see, thought experiments based on the principle of reverse anthropomorphism provide the basis for most studies in motivation analysis.

This voyage into the minds of sentient minds is going to be quite a journey. The nature of sentience is far too complex to be encapsulated within a one-line definition, such as 'the capacity to experience feelings'. Chapter 2 examines in detail the meaning and nature of consciousness and the sentient mind within the animal kingdom. To keep this enquiry as simple as possible, I shall consider the animal mind almost entirely as an abstract concept, within the brain and powered by the brain (mostly), but as an intangible compendium of information bank, instruction manual, filter and digital processor of incoming sensations and information. It is not too far-fetched to make the analogy with the digital computer and describe the brain as the hardware and the mind as the software. The neurophysiology involved in driving the hardware has its own beauty, but that is another story.

Through evolution by natural selection, animals have acquired behavioural skills appropriate to their design (phenotype) and natural environment. All animals are equipped at birth with a basic set of mental software: instructions genetically coded as a result of generations of adaptation to the physical and social challenges of the environments in which they evolved. This, which I shall hereafter refer to as their mental birth-right, is instinctive and hard wired. In some species that we may define as primitive, their responses to stimuli may always be restricted to invariant, hard-wired, preprogrammed responses to sensations induced by environmental stimuli. According to one's definition, this alone may be sufficient to classify them as sentient. However, throughout the animal kingdom, from the octopus to the great apes, we find overwhelming evidence of species that exhibit sentience to a higher degree. They build on this instinctive birthright and develop their minds. They learn to recognise, interpret and memorise new experiences in the form of feelings, good, bad or indifferent, and develop patterns of behaviour designed to promote their wellbeing measured, in all cases, in terms of primitive needs such as the relief of hunger and pain and, within the deeper, inner circles of sentience, feelings of companionship, comfort and joy. The ability to operate on the basis of knowledge acquired from experience, rather than pure instinct, enriches the physical and mental skills the sentient animal can recruit to cope with the challenges of life and promote an emotional sense of wellbeing. It also carries the potential for suffering when coping becomes too difficult.

The physical and mental skills and resources present at birth are those acquired through adaptation of their ancestors to the ancestral environment, because these were the skills that mattered the most. Animals that demonstrate deeper degrees of sentience have the capacity to develop these inborn, instinctive skills throughout their lifetime and teach these new skills to subsequent generations. Differing demands of differing environments mean that each species exhibits a portfolio of skills most appropriate to their special needs. It follows that, in our eyes, individual species may appear to be brilliant at some things and dumb at others. Raptor birds that hunt by day develop an exquisite visual ability to locate their prey whereas bats that hunt at night use radar based on ultrasound. The albatross can navigate its way home to its nest across the barren expanses of the Southern Ocean but will fail to recognise its chick if it has blown out of the nest. Domestication distorts the process of natural selection in two ways. We compel these animals to adapt to an environment largely determined by us, and this may be very different from that of their ancestors. We also introduce the entirely unnatural business of breeding: we tinker with the physical and mental phenotype of our animals to suit our needs for food, fashion, recreation or unqualified love.

We cannot observe animals through our eyes and conclude that any one species is better, or more highly developed than another. Each species adapts to meet its own special needs and the skills required to meet these needs vary in their nature and complexity. Pigs are good at being pigs, sheep are good at being sheep. Rats are very good at being rats because they have had to develop the physical and mental skills necessary for survival in a complex and frequently hostile environment. Sharks are very good at being sharks but, because they have thrived for millennia in a food-rich, stable environment, they have never really had to think. Many dogs are not very good at being dogs because they have not had the chance to grow up in an environment of dogs.

Human Attitudes to Animals

Most of this book is devoted to an exploration of the minds of sentient animals, their feelings, thoughts and motivation to behaviour seen so far as possible, through their own eyes. Human attitudes to animals would be irrelevant were it not for the fact that our actions, based on our attitudes, can have such a profound effect on their lives. In an earlier book, 'Animal Welfare: A Cool Eye towards Eden' (76) I wrote '*Man has dominion over the animals whether we like it or not. Wherever we share space on the planet, and this includes all but the most inaccessible regions of land, sea and air, it is we that determine where and how they shall live. We may elect to put a battery hen in a cage or establish a game reserve to protect the tiger but in each case the decision is ours, not theirs. We make a pet of the hamster but poison the rat. These human decisions are driven by the same incentives that motivate non-human animals since they reflect the will of us as individuals and as a species to survive and achieve a sense of well-being. We need good food and we seek highly nutritious eggs at little cost. We need good hygiene and seek to remove rats that carry germs. We choose to provide for our pets in sickness and in health because they enrich the lives of us and our children. We admire the tiger not only for its fearful symmetry but as a symbol of freedom itself, so we offer it more freedom than we give the laying hen. However, in either case it is impossible to escape the conclusion that both are living on our terms.*'

The history of human attitudes to animals (and to other humans) is awash with ignorance and inhumanity. The European Judeo-Christian belief was inscribed in Genesis as '*every beast of the earth and every fowl of the air. . .I have given for meat*'. The attitude of other religions to non-human animals varies. Of the Eastern religions, Taoism and Buddhism recognise the sentience of our fellow mortals and treat them with respect. More of this later. So far as I can gather, Confucianism regards non-human animals as commodities or tools, and therefore 'off the page' so far as philosophy is concerned. Islam and Judaism display rituals of respect for their food animals at the point of slaughter but these bring no comfort to the conscious animal while it bleeds to death. The Hindu veneration of the Holy Cow is driven more by fear of divine retribution than any concern for animal welfare.

The French philosopher Rene Descartes (1596–1650) sought to justify the Judeo-Christian attitude by asserting that humans are fundamentally different from all other animals because we alone possess mind, or consciousness. His notorious phrase *Cogito ergo sum* – I think, therefore I am – further implied *non cogitant ergo non sunt* – they don't think therefore they aren't. He saw non-human animals as automata, equivalent to clockwork toys, and thereby provided an 'ethical' basis for treating them simply as commodities on the assumption that it is not possible to be cruel to animals because they lack the capacity to suffer. His view may appear to us as totally lacking in any understanding of animals. However, he was not alone. For most of history, the moral concepts of right and wrong were applied only to intentions and actions within the human species. The utilitarian, Jeremy Bentham (1748–1832) was an exception when he wrote of animals 'the question is not can they reason. . .. *but can they suffer?*'. The supreme challenge to this limited concept of morality came from Albert Schweitzer who wrote '*the great fault of all ethics hitherto has been that they believed themselves to have to deal only with the relations of man to man. In reality, the question is what is his attitude to the world and all that comes within his reach*'. This became the basis for his principle of reverence for life (10).

The last Century has seen a steady progression of the evolution of morality into law. The UK Protection of Animals Act (1911) made it an offence to '*cause unnecessary suffering by doing or omitting to do any act*' (59). The 1997 Treaty of Amsterdam acknowledged that '*since animals are sentient beings, members should pay full regard to the welfare requirements of animals*' (73). The UK Animal Welfare Act (2006) imposed a duty of care on responsible persons to provide for the basic needs of their animals (both farmed animals and pets) (25). This act signified a considerable advance, since it is no longer necessary to prove that suffering has occurred, it is only necessary to establish that animals are being kept or being bred in such a way that is liable to cause suffering. These proscriptive laws are written in broad terms, which gives them the flexibility to deal with a range of specific circumstances. However, they beg several questions: 'what constitutes suffering, especially *necessary* suffering? 'what *are* the welfare requirement of animals?', and (above all) 'what is meant by sentience?' One of the main aims of this book is to guide all those directly and indirectly involved in matters of animal welfare (which means almost everybody) towards a deeper understanding of the complex biological and psychological properties of animal minds that determine their perception and their behaviour, thus determining the principles that should govern our approach to their welfare.

Despite the evidence of progress in the law relating to the protection of animals, there is still too much evidence of cruelty, both deliberate and mindless. Deliberate cruelty is a crime punishable by law and relatively rare. Mindless cruelty is far more common. It reflects a mindset conditioned by ignorance or training to the assumption that animals are automata, thus incapable of suffering. We are constantly presented with images of abuses to animals from all over the world. I cite only three examples.

A few years ago, Compassion in World Farming (CIWF) released a shocking video of behaviour in a small abattoir. Lambs for slaughter were hung up by driving a hook through their legs behind the Achilles tendon prior to stunning and having their throats cut. In this video, four lambs were hung on hooks and left to struggle while the slaughterman went off to smoke a cigarette. From the lambs' perspective, this was cruelty in the extreme. I suggest, however, that from a human perspective this may not have been deliberate cruelty but an extreme case of mindlessness. It had never occurred to him, or been explained to him, that sentient animals are capable of suffering. If he had been really cruel, he would have watched.

My most extreme personal experience of the mindless ill-treatment of animals came from a large commercial pig abattoir in Beijing. Pigs transported to the abattoir in crates had been gaffed by the neck and hauled out of their crates on long poles like inert sacks of corn. This was not only appallingly cruel, to our eyes, but spectacularly counterproductive because the pigs fought them every inch of the way. The Bristol team designed a humane handling system whereby the pigs were able to move out of the vehicles and down a well-designed passage at their own speed with minimal stress and human interference. The abattoir owners were delighted with this new system because they were able to reduce the number of staff needed to 'handle' the pigs by over 50%.

These two instances of mindless ill-treatment may be attributed to ignorance. However, ill-treatment on an industrial scale, carried out with the approval of the highest authorities, remains a problem in the so-called developed world and to the present day. The number of chickens killed and consumed by humans *every day* is approximately 70 million. Furthermore, most of them are unlikely to experience much that could be quality of life before they die. In the words of Ruth Harrison, the godmother of the Animal Welfare movement: '*If one person is unkind to one animal, it is considered as cruelty but when a lot of people are unkind to a lot of animals, especially in the name of commerce, the cruelty is condoned and, once large sums of money are at stake, will be defended to the last by otherwise intelligent people*' (29). It was Ruth who pointed out the absurdity of the UK Protection of Birds Act (1964) which required any caged bird to be given enough space to flap its wings but then stated *'provided this subsection does not apply to poultry'*. This subsection meant that, at the time, the Act did not apply to about 99% of caged birds. This is perhaps the most egregious example of the fallacy of classifying animals as commodities in term of their utility to us, rather than as sentient beings whose minds have been shaped by their genetic inheritance and their individual experience of life. It was sustained public pressure generated by pioneers like Ruth Harrison that compelled the European Union to pronounce in the Treaty of Amsterdam that '*Members shall, since animals are sentient beings, pay full regard to the welfare requirements of animals*' (73). This is a clumsy sentence from a clumsy clause that is also littered with caveats and exceptions for regional and religious practices. Nevertheless, it did recognise in law the principle that animals used by us for food, scientific enquiry, or health and safety legislation should not be considered simply as commodities but treated with respect and concern for their wellbeing.

Animal Behaviour Science

This exploration of the minds of sentient animals draws heavily on scientific studies of animal psychology and behaviour. The scientific investigation of animal behaviour is concentrated on two main themes. The first is the study of how animals behave in their natural habitat. This can establish their behavioural needs and the actions they perform to meet these needs. From this, we can build up a reasonably comprehensive picture of the resources (e.g. diet, physical and social environment) they require to achieve a sense of physical and mental wellbeing. With this information to hand, we can devise management policies that seek to address these needs whenever we modify their natural habitat to suit our own needs for food, companionship, sport, safety or scientific endeavour.

The second approach is to present animals with a set of questions relating to their perceived needs and measure their responses. This is the science of motivation analysis (16). The simplest version of this approach is the *Preference Test*. In a typical experiment, the animal is given a choice, e.g. between two foods or two environments and invited to demonstrate a preference. The choice may be between options that we guess might create more or less satisfaction (e.g. two types of bedding material for pigs), or between options that may be more or less aversive (e.g. barren vs. enriched cages for hens). One classic approach is to place the animal in a T maze that allows it to choose between the two options of taking the path to the right or the left. This can tell us quite a lot. Pet food manufacturers may discover flavours preferred by cats (although cats are fickle creatures). Designers of enriched environments for intensively reared pigs or chickens can get some idea of the fixtures and fittings that these animals appear to favour or avoid. However, preference tests can sometimes reveal evidence to indicate that the scientist and the experimental animals are not thinking the same way. In one such experiment, mice were asked to choose between two environments deemed by the scientist to be more or less enriched by traversing a narrow tunnel between the two. Most mice chose to spend the majority of time in the tunnel. For them, this was better than either of the choices on offer (66). The scientists had assumed the mice would choose on the basis of comfort, whereas, in their minds, we must assume that the primary need was for a sense of security. The scientists posed a specific question to these mice and got an unexpected answer. It was the wrong question, but they had a better understanding of mice as a result.

The main limitation of the preference test is that it makes no distinction between choices that are trivial and those that really matter. A more advanced approach to motivation analysis is to measure the strength of motivation by how hard an animal is prepared to work to get a reward in the form of a pleasant experience such as food, or relief from an unpleasant experience such as cold, pain, isolation, or a barren environment (16,45). Examples of the currency used to measure cost include the number of times the animal has to press a lever, or the amount of pressure it has to exert on a gate to obtain the reward. Specific rewards are ranked as more or less price elastic or price inelastic. Most animals, unless satiated, will continue to work for a food reward as the price is increased, which makes it price inelastic. The marginal reward of a different lying surface, e.g. straw vs. wood shavings may be price elastic: i.e. not worth too much effort. While the preference test can do no more than establish behavioural priorities, motivation analysis can determine how much these things matter.

The aim of motivation analysis is to devise tests that enable an animal (e.g. a rat or chicken) to demonstrate, by way of its actions, how it feels about the challenge with which it is faced, positive, negative, or indifferent. Having demonstrated that the test animal is motivated to act to receive a specific reward such as food or avoid a potentially unpleasant experience such as isolation or confinement, the scientist then measures the price the animal is prepared to pay to improve its welfare. They observe this behaviour, review the results in the light of current understanding as already described in the scientific 'literature' and form conclusions based on evidence as to the preferences and strength of motivation of the animal. This will be set out for publication in words, tables and diagrams. The scientist has used the medium of language to describe conclusions and decisions that arose first in the mind of the rat or chicken in order that other humans might better understand how it feels to be that chicken. This is reverse anthropomorphism, pure and simple.

There is another profound conclusion to be drawn from studies such as these; one that is key to our understanding of the minds of our fellow mortals. Presented with a specific question, which can be quite complex, the rat or chicken has analysed the problem, worked out a satisfactory response and memorised the actions necessary to achieve that response without recourse to the uniquely human medium of the spoken and written language. Moreover, as we shall see later, the ability to solve simple problems set by scientists in the laboratory can be a very limited measure of an animal's mental capacity. It pales into insignificance when set, for example, alongside the detailed large-scale maps that a pigeon needs to carry in its head if it is to navigate its way home. Animals with sentient minds have the ability to acquire and retain a great deal of knowledge and understanding without the need for language as we understand it nor reference to external banks of information stored in libraries and/or Google. What is more, these animals may be able to convey this understanding to their offspring, i.e. to engage in the process of education. We are only just beginning to understand the capacity of non-human animals to develop thought without language and convey these thoughts to others, but it is an ability worthy of the greatest respect.

Rules of Engagement

Two main themes run throughout this exploration of the minds of sentient animals.

Theme 1: *The needs of a sentient animal are defined entirely by its own physical and emotional phenotype, its environment and its education, and these are independent of our own definition of the animals as:*

Wild: subsets, game, (e.g. fox) vermin (rat), protected (badger)

Domestic: subsets, pet (dog), farm (pig), sport (horse)

In 'A Cool Eye towards Eden' I illustrated this theme with a picture of a brown rat in a larder. (Figure 1.1). I wrote at the time: '*A normal reaction to the brief glimpse of a rat in one's larder would be horror or, at least, a cold resolve to destroy the rat as quickly as possible, together with any others who happen to be around. Now study the picture more carefully. The rat is not only sleek to the point of being chubby but completely unalarmed by the flash photography, totally at ease in human company and altogether charming. Her name is Cordelia*'. Once we give the rat a name we provoke a shift in attitude. Nevertheless, Cordelia was a rat, and a rat is a rat, whether we classify it as laboratory animal, vermin or pet. She adapted wonderfully well to an enriched environment with loving human contact (my adult daughter, also an academic). If she had grown up in the company of other rats in the wild, she would have adapted equally

Figure 1.1 Cordelia at play. (from Webster, 1994)

well to that and, in the interests of her own survival, become fearful and dangerous in the presence of humans. If she had spent most of her life isolated in a barren laboratory cage, she would have had limited opportunity to develop her mind through lack of experience and thus be unable to handle complex decisions such as how to reconcile fear and curiosity in the presence of a novel stimulus. However, the essence of the rat mind is the same, whatever its circumstances. We have no right to assume that some rats are more equal than others. The behavioural and emotional needs of any sentient animal are determined by its own sentience, and these are entirely independent of our perception of its lovability, palatability, utility or nuisance value. In the case of wild animals, be they rats, badgers or, indeed, elephants, there are valid reasons ranging from human health to sustainable management of habitat to operate a form of population control. However, the principle of respect for all life directs that this should be as humane as possible. Where there is no clear need for population control, the policy for wild animals should be to leave them well and leave them alone. The most humane approach to the sensitive and sustainable management of wild animals is to preserve their natural habitat and stay out of their way.

Theme 2: *It is an anthropocentric fallacy to assume that the greater the similarity of an animal species to the human species, the more intelligent they are and the more worthy they are of our concern and respect.*

It is in our human nature to express most concern for the animals that look and appear to behave most like us. We are conditioned to believe that humans are the most intelligent of the animal species, so assume that animals that evolved in ways most similar to us must rank second. Thus, not only in popular opinion but also in legislation we give more rights to primates than to pigs. The anthropocentric fallacy was well recognised by Darwin and is implicit in the title of his seminal work 'The *Descent* (not the

Ascent) of Man'. To give just one illustration of the flaw in this argument, corvid birds (e.g. crows) are better at problem solving than chimpanzees. Much more of this anon. However, this argument based simply on the basis of problem-solving skills is, like all arguments based on selected evidence, far too simplistic. I shall seek to persuade you that it is pointless to claim that one animal species is more intelligent than another. Each sentient animal is born with, and further develops the mindset and skills most appropriate to its needs and these needs are defined by the environment to which it must adapt. When we seek to measure the intelligence of animals according to criteria that we humans would define as measures of intelligence, such as the ability to associate symbols with boxes that contain food rewards, we may conclude that the most advanced of non-human animals can just about match the intelligence of a three-year-old child. When we start to wonder about the skills that animals display in relation to things that matter to them, but which we cannot measure in the laboratory, like navigating the world, we can only conclude that, in some respects, their skills may be superhuman. These two themes crystallise into one single, central message. Our respect for, and actions towards, all species of sentient animals should be based on our best possible understanding of their life as they see it, not as we see it. In matters of human respect for animals, the question '*What is this animal for?*' has no meaning.

The essence of this book is an exploration of animal sentience: how it is determined by, and how it adapts to the physical and mental challenges of the specific environments to which they are exposed. Part 1, The Sentient Mind, skills and strategies, first explores the nature of sentience itself, how animals are motivated primarily by their feelings and the implications this has for their survival, success and wellbeing. It then examines the special senses, vision, hearing and olfaction, and the capacity of the mind to construct mental formulations based on information provided by the special senses and, from this, acquire knowledge and understanding.

Part 2: Shaping sentient minds: adaptation to the environment, examines the minds and skills of animals in groups defined not by their taxonomy, or their 'utility' (e.g. pet, farm, game, vermin) but by their habitats and the special challenges they present. I first examine animals in the natural environments of the waters and the air; least subject to interference from that most invasive of terrestrial species, mankind and therefore with most freedom to look after themselves. Terrestrial mammals are grouped according to their habitat. Animals of the savannah and open plains include the large number of herbivores (both wild and domesticated) and the smaller number of carnivores who prey on them. The chapter on animals of the forest gives special attention to the physical and social skills needed for life in the three dimensions of the tree canopy (Chapter 9). The last of the 'environmental' chapters (Chapter 10) considers our close neighbours, animals whose natural lives have been most affected by human interference, especially dogs, horses and animals confined to the farmyard or animal factory. In this section, I explore ways in which sentient animals build on the physical and mental tools acquired by way of their birthright in order to meet the special circumstances of their environment. These range from skills needed to manage primitive emotions like hunger, pain and fear to high-level cognitive formulations such as education and navigation, high-level emotional formulations like pleasure and grief, and the social graces of cooperation and compassion. Throughout this 'environmental' section, human attitudes are kept, wherever possible, off the page. The final section, 'Nature's Social Union', addresses the critical second clause in my title 'Why it matters' on the sound basis that sentient animals have feelings that matter to them, so they should also matter to us. This section examines human attitudes and actions to animals so far as possible through their eyes, not ours and reviews how we can apply our understanding of the sentient mind to meet our duty of care. I pose and seek to address questions such as: 'What can we learn from the animals that will help us to improve their lives and ours? How should we use this knowledge and understanding in the context of our responsibilities to our fellow mortals in the home, on farms, in zoos, laboratories and in the wild? Humans are burdened with the responsibility of care for the living world, based on the principle of respect for all life. This applies not only to animals in our direct care but to those whose lives we affect indirectly through our choice of diet or our competition for habitat (which means, just about all of them and all of us). Our aim must be to seek an honourable social union that achieves justice through proper respect to the things that matter to us and those that matter to them.

Wherever possible, my conclusions and assertions have been drawn from the evidence of science and the careful observations of those with sound practical experience of animal life. However, this can never be enough. I am just as concerned about what we don't know about animal minds as what we do. I shall often enter the realms of pure, although rational, speculation and I shall leave a lot of questions unanswered. This a brief exploratory voyage into largely unknown waters and makes no claim to be definitive. The subject is wide open. My observations, thoughts and ideas are offered as substance for reflection, discussion and an outline chart for future explorers.

Sentience and the Sentient Mind

'When I use a word' Humpty Dumpty said in a rather scornful tone, 'It means just what I choose it to mean, neither more nor less'

Lewis Carroll *'Through the looking glass'*

It is a truth generally acknowledged in morality and in law that we should regard the animals for whom we hold the responsibility of care as sentient creatures and treat them accordingly. This is easy to say but it poses a set of questions that don't have simple answers.

What, indeed, is animal sentience?
Is animal sentience an either/or thing or are there degrees of sentience?
If there are degrees of sentience, at what degree does quality of life matter to the animal (and so to us)?
What, if anything, is the difference between sentience and consciousness?
What do we mean by the mind?

As I pointed out in Chapter 1, we are faced with the problem of using words to describe the thoughts and feelings of animals that can experience and express these things perfectly well without them. We must therefore be absolutely clear that our

Animal Welfare: Understanding Sentient Minds and Why it Matters, First Edition. John Webster.

choice of word accurately reflects our interpretation of the thought or feeling that we are attempting to describe. We tend to use words like stress, sentience and consciousness in a very loose way and this can lead to confusion. Moreover, different interpretations of the meaning of words like sentience and consciousness can lead to fruitless dissent among scientists and philosophers who ought to know better, especially when we select words to describe the minds of animals for whom words have no meaning. I quote Wittgenstein who wrote '*philosophy is a battle against the bewitchment of our intelligence by means of language*' (85). Lewis Carroll, quoted above, was implying much the same thing but he is more fun. I shall therefore adopt the approach of Humpty Dumpty and use words like sentience and consciousness to mean just what I choose them to mean. Some may argue with my definitions, but they will be precise.

Sentience, Consciousness and the Mind

Search for animal sentience on Wikipedia and you are directed to Animal Consciousness, or state of self-awareness in a non-human animal. It proceeds to define consciousness in humans as '*sentience, awareness, subjectivity, the ability to experience or feel, wakefulness, having a sense of self and the executive control of the mind.*' This summary is correct in so far as the word consciousness is used to describe any and all of these properties (i.e. it can mean what you choose it to mean) but it does not begin to address the questions posed above. In particular, it fails to address the obvious variation in the nature of sentience within the animal kingdom, and how this might affect the expression of our reverence for life in terms of our actions in regard to, for example, a worm and an elephant.

One way to distinguish between the concepts of sentience and consciousness is to use sentience to describe sensations and emotions generated by stimuli ranging from the primitive (hunger, injury) to the complex (hope, despair, love, hate) and consciousness to describe the more or less complex cognitive processing of incoming information in the light of past experience (which may or may not involve feelings). Moreover, neither the words sentience or consciousness address the way in which sensations, information and emotions are interpreted by different animals and the impact these things may have upon their welfare. We need a far more comprehensive analysis of the nature of sentience itself and the operation of the sentient mind.

The Five Skandhas of Sentience

The most satisfactory *scientific* exposition of the varied nature of sentience within the animal kingdom that I have discovered in my reading is contained within Buddhist philosophy (65). This recognises five categories: *'skandhas'* which are present to a greater or lesser degree in all living creatures. It recognises all animals as sentient but some more sentient than others. The five skandhas of sentience are matter, sensation, perception, mental formulation and consciousness. These are illustrated in Figure 2.1 as five concentric circles of increasing depth, signifying increasing complexity from the outer, superficial circle of matter to the deepest circle of

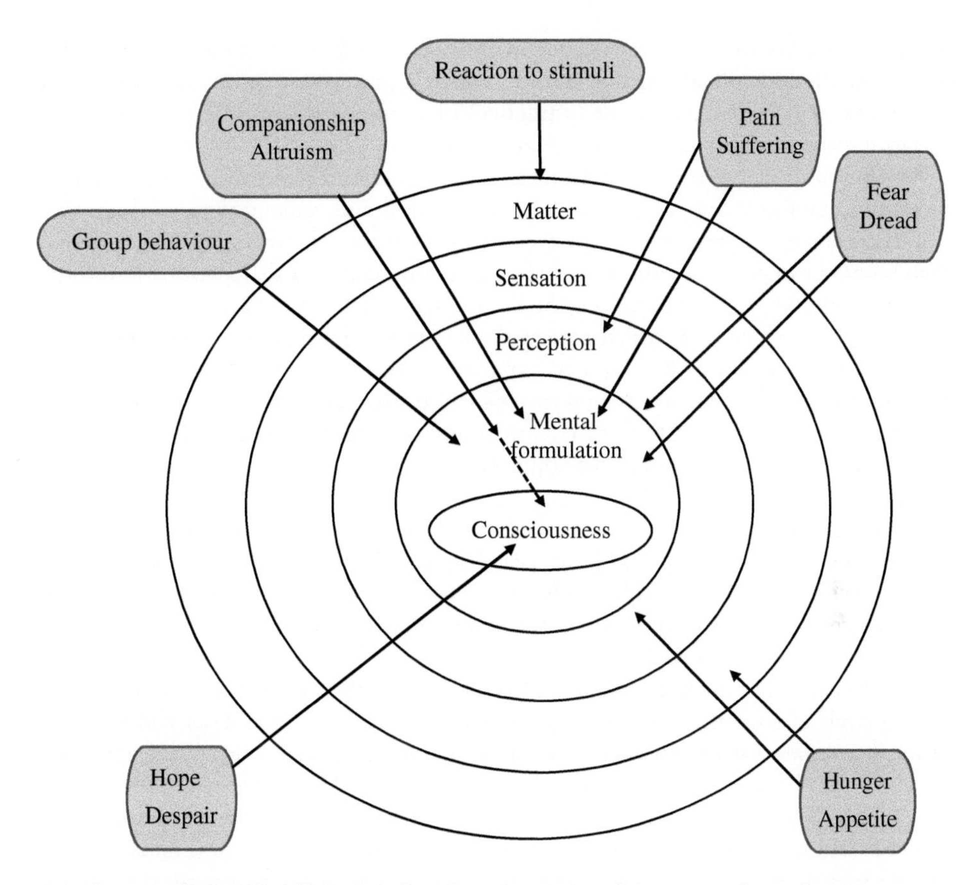

Figure 2.1 The five skandhas or circles of sentience. The solid arrows indicate the known extent of sentience involved in different forms of experience and social behaviour. Dotted lines indicate possible but unproven extension of sentience into the inner circles. The full interpretation of this diagram is provided in the text.

consciousness. Figure 2.1 also presents estimates, based on evidence relating to animal behaviour and motivation, of the degrees of sentience involved in the interpretation of primitive sensations such as hunger and pain and expressions of more complex behaviours and emotions such as companionship, altruism, hope and despair. These will be considered in due course.

Matter describes living organisms as defined by their physical structure and chemical composition and the chemical and physical processes that enable them to operate within a complex environment. This category embraces all plants and animals. It includes the ability to react to environmental stimuli, like the movement of sunflowers towards the sun, or the movement of amoebae away from an acid solution, without necessarily involving sensation as we would define it.

Sensation describes the ability of living creatures to experience feelings, and the intensity of feelings that take them out of their comfort zone. These include pain, severe heat and cold, hunger and fear. We may, with some confidence, assume that this property is restricted to animals. At this level of sentience, animals interpret these sensations as unpleasant (aversive), pleasant (attractive) or unimportant (indifferent) and these sensations will motivate them to take action, if necessary, to adjust how they feel.

Perception describes the ability to register, recognise and remember objects, experiences and emotions. Simple examples of recognition and remembrance include 'this food is good to eat', 'this animal (not necessarily identified as my mother) is the animal from whom I can get milk', 'this electric fence caused me pain, I shall avoid it in future'. Animals with the property of perception do not just react to stimuli as they occur; they can learn from experience. which carries the important message that they do not just live in the present. This enhances their capacity to cope with the challenges of life but increases the potential for suffering if the challenges are too severe, too prolonged, or if they are in an environment that restricts their ability to perform coping behaviour.

Mental formulation describes the ability to create mental pictures (or diagrams) that integrate and interpret complex experiences, sensations and emotions. This improves the ability of animals to learn from experience by gaining some understanding of the mechanisms of cause and effect. This increases their capacity to cope with challenge but equally, increases the potential for suffering if they find themselves unable to cope. The ability to create mental pictures also creates the capacity to develop the mind through education, given and received.

Consciousness: In the Buddhist skandhas, the word consciousness is restricted to the deepest circle of sentience and equates to human consciousness, best described as 'being aware that we are aware'. This carries the potential for advanced forms of social behaviour, both good and bad, such as empathy, compassion and cheating.

The Schweitzer principle of reverence for life requires us to respect all degrees of sentience and this is entirely consistent with our new moral and practical imperative to practise planet husbandry: to sustain and conserve the balance of nature for the welfare of all life. It does not, however, compel us to apply the same set of rules to a dandelion as to a horse. Animals whose degree of sentience extends only to the property of sensation will respond to primitive sensations such as pain, malaise, hunger and sex in a way that may be intense and probably adaptive but, by this definition, is hard-wired and does not necessarily involve what we might understand as emotion. However, the UK Scientific Procedures Act 1986 (24) recognises that the property of sensation is sufficient to give animals protected status with regard to actions likely to cause pain, suffering, distress or lasting harm, and requires these actions to be set against possible benefits to society. Species given protected status by the Act currently include all vertebrates and the invertebrate cephalopods. In the light of new research, this may have to be extended to other invertebrates. I shall have more to say on this later. In a broader moral context, it accepts that a primitive sensation such as pain may feel the same to a fish as to a dog.

It is, at this point, necessary to make a distinction between the possession of sentience as defined by the Buddhist skandhas and recognised by the UK Scientific Procedures Act, and the possession of a sentient mind. Nearly all the issues raised in this book relate to non-human animals that demonstrate properties of perception, mental formation and consciousness that enable them to develop their minds in ways designed, where possible, to promote their wellbeing. Hereafter, when discussing properties exclusive to these species, I shall refer to 'deep sentience' or the (sentient) mind.

Understanding the Sentient Mind

The mind, as such, does not actually exist. It is a complex abstraction that we can never fully grasp. The brain exists, which makes it easier for scientists to look into it using expensive bits of kit. We can observe the brain and its operation in ever increasing detail and record the electrical and chemical transmission of information. We can link specific sites and operations in the brain with specific thoughts and actions and provoke these actions with externally or internally applied stimuli. We know a lot about the biochemistry of emotion and the interactions between emotion and cognition: ways in which how we feel can affect the way we think. All this is fascinating stuff and vital to our understanding of the diagnosis and treatment of mental disorders but, so far, it falls a long way short of creating a coherent explanation of what we imagine to be the mind.

The question 'What is mind?' has engaged, employed and escaped the grasp of philosophers and scientists for more than 2000 years. Most of us adopt a more relaxed approach encapsulated by the exquisite phrase 'be philosophical, try not to think about it'. We take it for granted that mind is a useful word to describe how we each interpret the world the way we see it. We might call this our mental state, but this is, once again, a circular argument. The majority position today, shared by philosophers and scientists is for *physicalism*, the assertion that mind is not an abstraction: everything in the mind is physical. Reductive physicalists assert that all mental states will eventually be explained in terms of measurable neurophysiological processes. This approach has obvious appeal to computer scientists developing better brains (and minds?) through artificial intelligence (AI). I am more sympathetic to the non-reductive physicalists who argue that while all of the mind may be in the brain somewhere, it will never be possible to provide a complete (or even useful) description of the mind entirely in terms of neurobiology. We shall always need both the psychological and the common sense, every day, approaches to our understanding of mind if we are to make any practical sense of it at all.

My wish is to understand the sentient mind, especially in non-human animals. I illustrate my approach with a simplistic but neurobiologically acceptable model of how a sentient animal interprets stimuli and sensations and what motivates it to respond (Figure 2.2). The control centre of the animal (the brain) is constantly being fed with information from the external and internal environment (outside and inside the body). Much information, e.g. our perception of how we stand and move in space, is processed at a subconscious level. In Figure 2.2, these actions are controlled by stimuli that pass directly from receptors in the central nervous system (brain and spinal cord) via the motor nerves to the muscles that control posture and movement. Having learned to walk, we are able to control our limbs without recourse to thought or emotion. Many animals, e.g. raptor birds and top cricketers, have an exquisite ability to process the movement of an object in flight and programme their own movements so as to catch it. This is an amazing property of the neuromuscular system, but the only 'conscious' element was the decision to go for the bird/ball in the first place.

The decision to act (or not) in response to a stimulus must involve some degree of interpretation. The brain of all sentient animals is equipped at birth with a foundation programme for survival constructed from the specific gene-coded information acquired by its ancestors through generations of adaptation to the challenges of the environment in which they evolved. I refer to this property of mind as their *mental birthright*. For

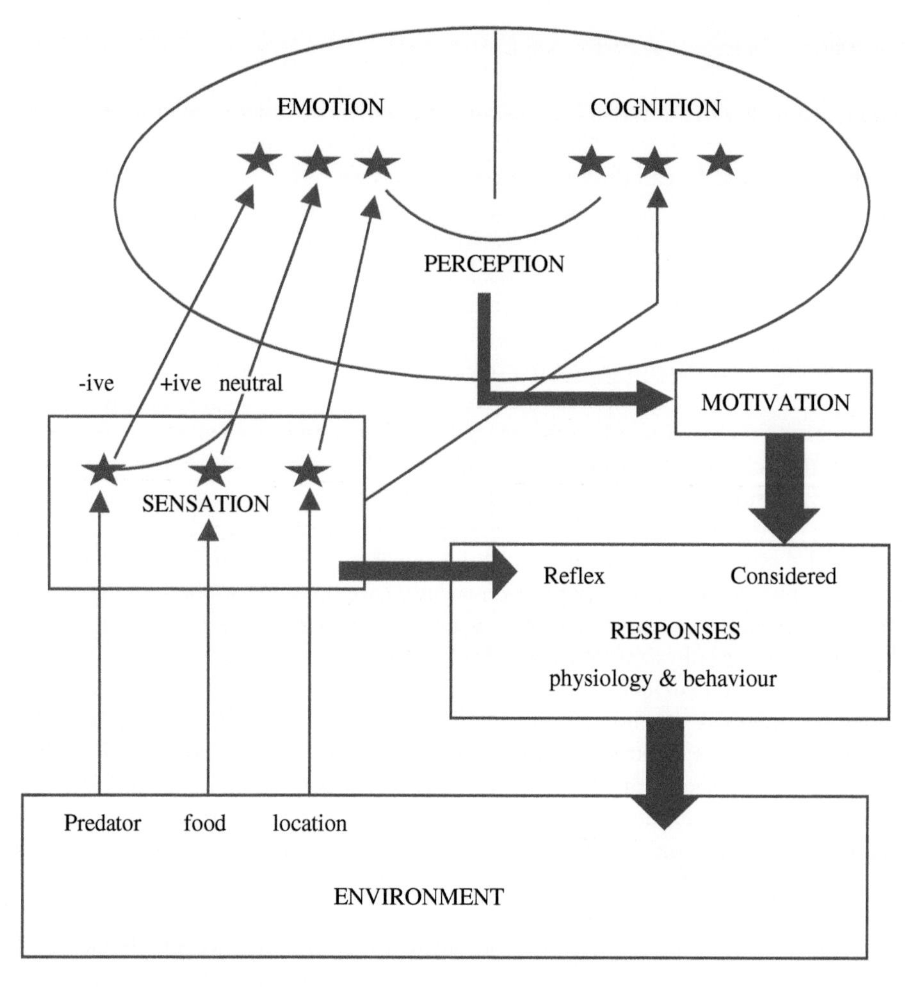

Figure 2.2 The sentient mind.

animals that demonstrate only the property of sensation, this may define the limits of their sentience. For animals that demonstrate the three inner circles of deep sentience, this foundation programme is expanded and enriched over time as a result of learning and experience. In animals with the power of perception, an incoming stimulus does not enter an empty room but becomes just another guest at the party, one that may or may not make an impact. In Figure 2.2, signals from external sensors such as the skin, the eye, the ear and internal sensors of, for example, biochemical parameters such as blood sugar, are recognised by the primary receptors as sensations (70). Much of this information from external sensors such as the skin and the eye will invoke a subconscious, reflex response like immediate withdrawal from a source of pain or threat. However. signals interpreted by the primary receptors as sensations will also pass to a second set of neural receptors linked to emotion (how it feels) and cognition (how it thinks). Negative stimuli like pain and hunger will be interpreted in the emotion centre as bad feelings and motivate the animal to conscious action to remove the source of pain or seek food. Other information, such as location or time of day may be interpreted simply as information with no emotional component and pass directly to the cognition centre. The response of the brain to incoming sensations and information may or may not involve a cognitive element but the motivation to conscious (as distinct from reflex) behaviour will nearly always have an emotional component. Indeed, I would argue that this is the essence of sentience.

To illustrate this concept, consider the primitive but clearly conscious sensation of hunger. The appetite control centre, which can be accurately located in the brain, will monitor internal stimuli such as low blood glucose or incoming sensory information from hunger pangs arising in the gut. It will also respond to natural external stimuli, such as the arrival of food, or a conditioning stimulus, such as the bell that Pavlov rang to condition dogs to anticipate the arrival of food. This direct information, like all incoming stimuli, will be interpreted in the context of existing food files in the brain carrying information from past experience as to the palatability, or possible poisonous nature of the food, or, if no food is currently on offer, the likely time to the next meal. If the animal is hungry and no food is available, it will experience a negative emotion that may range from slight to severe depending on how hungry it feels. When food arrives, it will feel good to a degree depending on the intensity of its former negative state (how hungry it felt) and the appeal of the food on offer. If the animal is already full up, it may treat the arrival of another meal with indifference: the information will be interpreted as neutral. My cats are masters of this studied indifference.

The notion that complex incoming information can be categorised on an emotional basis simply as positive, negative and neutral may seem overly naïve, but it is supported by classic experiments in neurobiology (33). Direct recordings from nerves in the brains of sheep reveal that when they are presented with food, or even pictures that indicate food (e.g. sacks of grain, bales of hay) these trigger signals in neurones that convey a positive message (a good feeling). Other images, e.g. pictures of dogs or humans, trigger a negative message (danger). However, when the sheep were shown an image of a familiar human carrying a bucket of food, these two categories of information (food and potential predator) were processed at the emotional level and passed on as a simple, unconfused positive message, 'I feel good'.

Animals may need only the first two circles of sentience to experience hunger as a sensation that increases in severity the longer they go without food, and so increases the strength of motivation to seek a meal. Animals that possess the third category of sentience, namely perception, can adjust their actions in the light of their past experience, choosing, for example, high energy foods that they associate with greater satisfaction in the past and avoiding foods that they remember made them feel sick after the meal. I shall have more to say about this when considering the special skills developed by different species to meet their special needs. The message at this stage is that an animal with the faculty of perception can remember the nature and intensity of sensations that induce positive and negative emotional states and will be motivated to actions designed to make it feel better at the time and avoid circumstances likely to make it feel bad in the future. At this level, sentience can truly be described as 'feelings that matter'.

In Figure 2.2, I have located the cognition and emotion centres on the left and right sides of the brain (we are looking at it from the front). Although this is overly naïve in an anatomical sense, it is consistent with current understanding of *lateralisation* within the conscious reception and control centres in the brains of vertebrates (61). In mammals, the cerebral cortex is divided into two distinct halves, with some linkage via the corpus callosum. Nearly all sensory and motor nerves cross over before entering and after leaving the brain so that the left side receives sensation from and controls muscles in the right side of the body and *vice versa*. So far, straightforward. The really interesting feature of lateralisation is that the left and right sides of the brain interpret incoming information and sensation in different ways. In humans, the left side of the brain appears to be primarily involved in the processing of information in detail, the right side with getting an overall feeling about the big picture. Humans with damage to the left brain have an impaired ability to understand and use language. Malfunctions in the right brain are associated with disorders of emotion: depression and delusions. All this reinforces the notion that the sentient mind possesses the properties of both emotion and cognition. Sometimes, it operates on the basis of feelings and thought, but always on the basis of feelings. This concept was central to the thinking of the psychiatrist Iain McGilchrist who entitled his investigation of the divided brain '*The Master and his Emissary*' (50), the master being the right, emotional side holding the big picture, the left being the well-trained civil servant with the learning and skills needed to do the right thing. I should add that Plato was there first with his allegory of the mind as a charioteer driving two horses, passion and logic.

The relevance of lateralisation to our understanding of the sentient mind in non-human animals is beautifully illustrated by the behaviour of birds, who have no corpus callosum so minimal links between the left and right brain. Vision, sensory input from the eyes, like nearly all other sensory information, crosses over *en route* to the brain. Thus, information from the right eye is processed in the left brain and *vice versa*. When chickens are foraging for food, seeking information, they tend to favour their right eye, so transmit specific information to the left side of the brain where it is interpreted in detail. When on the look-out for, or under threat from a predator, they favour the left eye, so stimulate a broad emotional response, fear, fight or flight (62).

To summarise the argument so far: the sentient mind operates primarily on the basis of feelings, modified to a varying degree by cognition, simply expressed as the ability to think about sensations and emotions. The power of cognition is widespread throughout the animal kingdom. We already know a lot about the cognitive abilities of mammals and birds and are discovering more and more evidence of the capacity of fish and invertebrates like cephalopods to construct complex mental formulations. I shall have much more to say about this. The point I wish to emphasise at this stage is that, for sentient animals, including humans, most of the time, cognition is acting in the service of emotion.

I make no claim that this is a novel insight. Two observations from David Hume, from his 1740 Treatise on Human Nature (230), are of note.:

Reason is, and ought only to be the slave of the passions, and can never pretend to be any other office than to serve and obey them.

The causes *of these passions are likewise much the same in beasts as in us, making a just allowance for our superior knowledge and understanding.*

Later philosophers (but not Darwin) have quarrelled with this bleak, animalistic view of nature. Humans may argue, without clear evidence, that we are better than this because we alone possess the fifth skandha, namely consciousness, the facility of being aware that we are aware. Clearly this has enabled some of us to plan and do clever things. However, we evolved as sentient creatures and the evidence strongly suggests that we too, most of the time, are motivated primarily by how we feel, whether our considered actions be selfish or altruistic. This is a good reason why I am confident that AI will never reproduce the human brain. For sure, it can make information processing computers that are better informed and quicker thinking than ours, so do amazing things such as creating better antibiotics, or playing better chess. However, I do not envisage an artificial brain driven by how it feels. Scientists will, I am sure, develop an artificial brain that *appears* to act according to how it feels but I suggest that this will be no more than a brain that acts according to how they feel it should feel, which is not the same thing.

Animals do not require the deepest circle of sentience, namely consciousness, in order to interpret and remember past experiences and the feelings they arouse. They cannot be said to live only in the present. Their emotional state will be defined by their expectations of the future in the light of past experience. This has profound implications for our understanding of the impact not only of primitive emotions such as pain and fear but also so-called 'higher' feelings such as hope and despair, comfort and joy.

Pain and Suffering

Pain in humans has been described as '*an unpleasant sensory and emotional experience associated with actual tissue damage*'. This is a good definition insofar as it recognises both the elements of sensation and emotion. Too many physiologists in the past have ducked questions concerning the emotional elements of pain, preferring simply to consider nociception, the sensation of pain. Some have argued that pain, even in humans, is such a subjective experience that it is not open to scientific explanation. To answer them, I again recruit Wittgenstein '*just try, in a real sense, to doubt someone else's fear and pain*' (85). I am happy to extend that doubt to other sentient species. In what follows, I shall adopt the simple model illustrated by Figure 2.2 and describe sensations and emotions associated with pain simply as positive or negative, in full recognition of the fact that these bald terms lack subtlety of meaning. I shall use 'mood' when I refer to emotional responses. Many scientists prefer the word 'affect'. I can think of no good reason why.

The possession of a sentient mind brings with it the capacity to suffer. This feeling may be modified by cognition. For example, the mood of a woman with severe abdominal pain will differ according to whether she knows she has cancer or is giving birth to a baby, but the sensation may be the same. We can say with some certainty that potential sources of suffering in non-human animals include pain, fear, hunger and thirst, severe heat and cold, malaise (feeling ill) and exhaustion. We can say with equal certainty that we humans can suffer from anxiety and depression, boredom and frustration, loneliness and loss. We have a duty to explore the extent to which we may share these 'higher' emotions with other sentient species.

We can describe our own experience of pain as a negative experience, a bad thing, both in terms of the immediate sensation and its emotional consequences. We seek to avoid sources of pain, we take drugs to reduce the sensation and, if the pain persists, our mood takes a turn for the worse. In recent years, there have been many excellent, compassionate studies of pain as a sensory and emotional experience with mammals, birds and fish that convince me of their capacity to suffer. My personal links with this largely involve the work of colleagues at the University of Bristol with broiler chickens (34,39). I shall briefly summarise this work as an illustration of a more general theme.

The word 'broiler' describes chickens that are selectively bred and reared to produce meat as quickly and cheaply as possible. Modern broiler strains can progress from the moment of hatching to the moment of slaughter in 40 days or less. As a result of this, extreme pressure for increased productivity of an increasing number of birds has developed what the trade has called, without shame, 'leg weakness'. As they approach slaughter weight the birds become increasingly reluctant to move, many go off their legs altogether and die or have to be culled, not least because they can no longer reach the feeders. The pathology that gives rise to leg weakness is complex. It can involve both abnormal bone growth and joint damage, often exacerbated by infection, but it is all precipitated by the fact that these birds are outgrowing their strength. There are serious issues of animal welfare and human ethics arising from the problem of leg weakness in broilers. These have been addressed at length elsewhere and it is fair to say that the industry, largely in response to consumer pressure, has gone some way to putting its house in order. However, this is outside my current brief, which is strictly concerned with the extent to which the pain associated with this condition of leg weakness strains of this bird that have been selected for rapid growth may be a source of pain and suffering.

To start from the extreme position of the devil's advocate: it would be possible, in the absence of further evidence, to claim that the progressive loss of mobility in fast-growing strains of broiler chickens could simply reflect increasing mechanical interference with joint movement, possibly accompanied by an increasing imbalance between muscle strength and body weight, both things not necessarily accompanied

Pain: How do We Know It Matters?

Immediate reaction

- Alarm and escape: conscious sensation but no evidence of emotional response

Modified behaviour

- Rest and locomotor changes: evidence only of adaptation to a negative experience
- Learned avoidance: evidence of memory of a negative emotional state
- Reduced positive behaviour (grooming, exploration, play): evidence of chronic change in mood

Altered mood

- Apathy, reduced appetite, reduced strength of motivation to respond to positive and negative stimuli: strong evidence for depression

Response to analgesics

- Improved locomotion after external administration only indicates relief at removal of painful sensation
- Self-selection: conscious, learned behaviour expressing powerful motivation to avoid a negative emotional state (feeling bad).

by pain. Others have argued that chickens may experience the sensation of pain unaccompanied by any element of emotional distress, and so not actually suffer. We cannot simply dismiss these arguments as heartless, we must construct a solid, evidence-based case if we are to show that chickens can experience pain and that this pain causes them to suffer.

Learned avoidance of a behaviour likely to cause pain, such as jumping down from a perch, provides some evidence for an emotional response. Changes of mood, such as reduced appetite, grooming, or positive actions such as exploration and play, offer further, but not yet totally convincing evidence that pain has an emotional element in the species under observation (here, the chicken). The most convincing evidence that animals suffer when in pain comes from experiments that involve the self-selection of analgesics. Elegant experiments have been undertaken in which rats, chickens and several other species of sentient animals have been given free choice to self-select analgesic (pain-killing) drugs in their food or water (39). In these trials, it is important to avoid drugs that induce euphoria, such as opiates, since all animals tended to favour and become addicted to these. With non-euphoric analgesics such as the non-steroidal anti-inflammatory drugs (NSAIDS), animals without damage likely to cause pain tended to prefer the food without the drug, probably because the treated food tasted strange. However, chickens (in this case) that displayed signs of abnormal locomotion preferentially chose the food containing analgesic and consumed enough of this food to administer what the manufacturers would advise to be the correct analgesic dose. They learn how to relieve their pain through self-medication. To my mind, this is the final, convincing link in the chain of evidence that proves that, in these species, pain is a matter of sensation and mood. We may safely conclude that when these animals are in pain they really suffer.

Fear and Dread

Fear is an essential, highly functional, primitive sensation that acts as a powerful motivator to behaviour designed, where possible, to avoid threat. It is also an educational experience since the memory of previous threats, the action taken in response to those threats and the consequences thereof ('was it less bad than I feared or worse?') will obviously affect how the animal feels next time around. Causes and consequences of fear are illustrated in Figure 2.3. This identifies three main threats, novelty, innate threats and learned threats. Neophobia, or the fear of novelty, is an obvious survival mechanism. Success in life depends on achieving the right balance between curiosity (to develop survival skills) and caution (to avoid danger). Among these skills is the ability to learn the distinction between threats that are real or imagined. Most innate fears (e.g. primates' fear of snakes) have survival value. Arachnophobia (human fear of spiders) would appear to be a lasting remnant of our primitive past. While these innate threats are relatively hard-wired (i.e. built into the mental birthright) they can be overcome by experience. Fear of a learned threat is self-evidently one that is acquired by experience, e.g. the fear shown by many dogs on visits to the vets. The dog that is returned to boarding kennels prior to the annual family holiday, or even watches the family packing up, may experience a more advanced form of learned fear: that of desertion by key members of its social group.

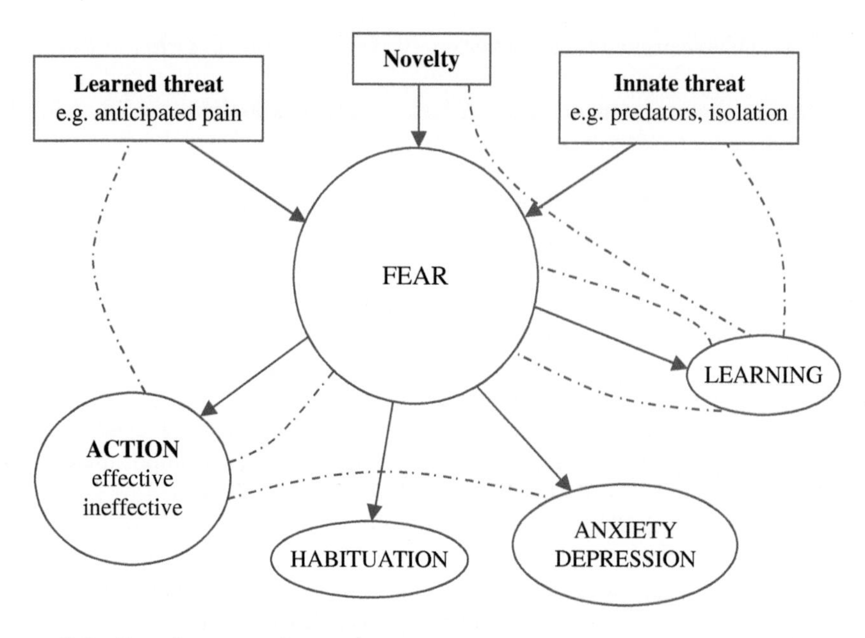

Figure 2.3 Fear, threats reactions and consequences.

Stimulated by fear, the animal takes what action it can. If it learns that its action has been successful, it will know what to do next time. It has discovered that it can cope and will experience an increased sense of security. If it discovers that it cannot resolve the problem either because its actions were ineffective or because it is in an environment that restricts its ability to act, then it may suffer a chronic, non-adaptive change in its emotional state, or mood, within a spectrum ranging from high anxiety to severe depression. In simple terms, acute fear morphs into chronic dread.

Coping with Challenge: Stress and Boredom

In our comfortable, sheltered world, we may be inclined to succumb to the delusion that all stress is a bad thing and that, if we have any animals in our care, we should seek to protect them from stress at all times. In the natural world, what we may term stress, but what is better defined as challenge, is a fact of life and learning to cope with stress is an essential part of adaptation to the environment. In the natural world where an animal has to look after itself, it will constantly be presented with a variety of challenges: the need to find food and water to avoid suffering from hunger or thirst, the need to avoid sources of pain and fear, the need to shelter from the threat of severe heat or cold and the need to achieve sufficient rest to avoid the prospect of exhaustion. Animals need to cope with these challenges and may suffer if the challenge is too intense, or if it is faced with too many conflicting challenges occurring at once.

Figure 2.4 provides a simple illustration of the concept of coping. The animal with a sentient mind normally operates within a sphere of adaptation to a variety of

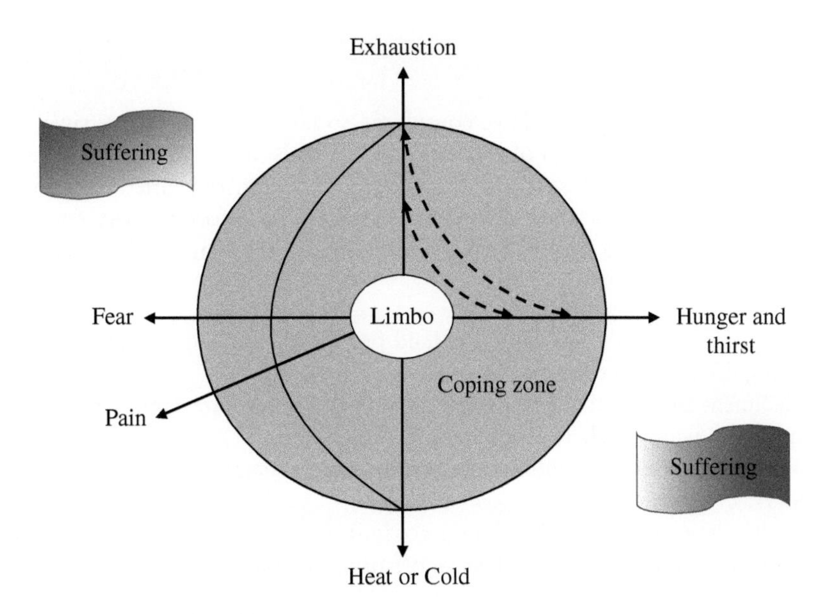

Figure 2.4 Coping with challenge.

challenges. The herbivore, for example, that needs to graze for many hours to obtain sufficient nutrition, also needs to rest. On a sparse arid pasture in the tropics, it may also be faced by conflict between the need to take in enough food and the need to avoid heat stress. Provided the challenges are not too great (the pasture is adequate, and the day is not too hot) it can adjust its position within the sphere of adaptation without too much difficulty: in short, it can cope. However, as the intensity and complexity of the challenges become more severe, it is driven ever closer to the perimeter of the sphere of adaptation and may pass through into the domain of suffering. This illustrates the point that stress and suffering are not the same thing, Suffering occurs when the animals fail to cope, or approach the point of failure.

At the centre of the sphere of adaptation is a zone that David MacFarland calls 'limbo', wherein there are no challenges at all (46). Since animals have evolved to manage challenge, this must be considered an unnatural state and most unlikely to occur in the wild. However, for animals under human control, such as a rabbit in a hutch, a bird or lion in a cage, a pregnant sow chained to the floor, it can too often be a fact of life. This raises the question: 'Can sentient animals suffer from boredom?' The answer is almost certainly yes. Rats, sows, horses, lions and elephants (to cite but a few examples) kept in barren environments develop stereotypical behaviour, typically expressed as purposeless, repetitive activities like pacing or bar chewing. Similar stereotypical behaviour (rocking and head banging) is seen in children confined in barren orphanages and for much the same reasons. The extent to which this behaviour is an expression of raw suffering or a mechanism to reduce suffering is open for dispute. What is not in dispute is that it is an indication that something is seriously wrong.

Social Life

The emotional sensations that I have considered so far have all been aversive: hunger, fear and pain, potential sources of suffering that animals act to avoid. The next step is to explore how animals with sentient minds seek to maintain and enrich their quality of life through their attention to positive stimuli: sources of comfort, security, companionship and pleasure. Expressions of the higher emotions in the context of the social life of animals include the following:

- Establishment of family bonds that extend beyond the immediate, instinctive needs of feeding offspring to the point of weaning or fledging.
- Establishment of special relationships outside the immediate family, reinforced by affiliative behaviour such as grooming.
- Formation of alliances within the group for assistance in encounters with outsiders.
- Unselfish expressions of empathy and compassion for others, such as consolation behaviour to one who may have lost an offspring or come off worse in a fight.

These expressions of feeling all require at least the fourth level of sentience, the ability to establish mental formulations. The extent of this expression in different species is determined to a great extent by their birthright, itself laid down by the genetic consequences of many generations of adaptation of their ancestors to the environment. Sheep and horses are herd animals, tigers are solitary, wolves and elephants live in extended family groups. However, all these varying patterns of social behaviour reflect the same motivational drives. On a population basis, successful behaviour is that which promotes genetic fitness, which is a hard-wired (mindless) aim. However, the genetic fitness of the population depends upon the behaviour of individuals, who are motivated not by an understanding of Darwin but by their immediate individual and family needs. These will always include primitive needs for security, sex and survival, but may, in many sentient animals, include 'higher' emotions such as friendship, empathy and compassion.

Most herbivores, being prey species, live in large herds. At the simplest level, the success of this behaviour does not require a degree of sentience deeper than that of hard-wired sensation. The probability of one sheep being killed by a passing wolf is obviously greater if it is out on its own than if it is within a large flock. However, there is good evidence for more highly developed flocking behaviour in herd animals. Musk oxen in the Arctic, with calves at foot, defend themselves against attack from a wolf pack by forming a circle, with the aggressively horned adults pointing outwards and the calves safely in the middle. The primary motivation for this behaviour is the hard-wired, instinctive need for survival of the group (not necessarily the individual) but its complexity indicates that this is a strategy that has benefited from social learning. Some herbivorous species, such as the Rocky Mountain goat, do not form large compact flocks. Their speed and agility in difficult terrain give them odds-on chances of escape. Predator species tend to be solitary, like the tiger, or operate in family groups, like wolves. Again, the proximate reason for this is obvious. The tiger needs no help to stalk and kill its prey; its greatest threat is from another tiger.

Wolves are successful when they seek and destroy in packs. Conspecifics outside the family groups are likely to be seen as competitors and a potential threat to survival and genetic fitness.

The primary motivation to social behaviour in a herding, prey species is the need to achieve security by reducing the risk from predation. However, animals compelled to live in close proximity to one another in large groups also need to establish a *modus vivendi* that ensures a quiet life. Most achieve this through the establishment of a hierarchy, or pecking order, which may involve aggression in the early stages but usually settles down unless the group is further disturbed, e.g. by new arrivals. In these groups, some animals may establish especially close contacts with chosen individuals, who may or may not be related to them, and reinforce these contacts with bonding behaviour such as social grooming. This does not necessarily imply friendship. It may just be a way of increasing strength in numbers when faced by aggressive neighbours. I shall have more to say about social behaviour in subsequent chapters. However, all the behaviour described so far, may be simply interpreted by the motivation to seek safety, security and a quiet life. In many species that depend for their individual survival on living in large groups, like shoals of fish and possibly sheep, the motivation to huddle for safety within a crowd is likely to be instinctive and may not require a greater depth of sentience than simple sensation. Those who learn to exploit the advantages to themselves and their families of living in groups, adapt their behaviour accordingly. This learned behaviour, which is seen in the social primates, must, at least involve the property of perception and, in many cases, the capacity for mental formulation.

Social contact, if only within the family, is necessary to ensure that animals get a good education. Sentient animals develop their life skills by learning from others. Once again, there are wide species differences in the need for education. The shark is able to look after itself from an early age without help. Fledgling birds are pretty helpless and need to be taught survival skills by their parents.

We would, I think, describe human friendship and love (with hate as the polar opposite) to be an expression of the inner circle of sentience, namely conscious expression of social behaviour. Evidence of friendship in animals is based on observations of behaviour that include strong affiliative bonds with special individuals, inside or outside the family, and grief at the death or departure of a comrade. Many animals display signs of distress after loss of contact with preferred individuals. Donkeys are particularly likely to suffer severe distress after the loss of a long-time friend and will not be consoled by contact with others. We have, I believe, sufficient evidence to include friendship within the scope of emotions that we should recognise and respect in at least some of the sentient animals. The motivational basis of these emotions is uncertain, even in humans. Many stern behaviourists argue that they can be explained simply in terms of the genetic imperative, classically expressed by J.B.S. Haldane who wrote '*I would be prepared to lay down my life for two of my children or eight of my second cousins*'. The grief expressed by a donkey, or an elephant at the death of a close companion may simply be an expression of personal loss. According to the Haldane argument, true love can only be expressed in terms of true altruism, behaviour designed to benefit an individual or group outside one's genetic pool despite the fact that it brings no benefit and may come at some cost to oneself. There is some evidence for true altruism in animal societies. Chimpanzees, bonobos and rooks(!) have been observed to offer comfort to others showing signs of distress. More of this anon.

Comfort and Joy

Any animal in a natural, challenging environment needs to look after itself and stay out of trouble. The most important essentials for survival and a sense of wellbeing are to establish a supply of food that exceeds the energy cost of acquiring it and learning to cope with stress. To achieve a sense of security, the animal must come to terms with the most important instinct for survival, namely *neophobia,* fear of the unknown. This requires a balance between curiosity and caution. Generally speaking, young animals have a high sense of curiosity. In nature, most young mammals, birds and a few species of fish explore the world under the close eye of at least one parent. If all goes well during this process, they learn to identify and distinguish between threats that are real and those that are imaginary. With increasing knowledge and maturity curiosity gives way to caution: adults are less inclined to explore and more likely to restrict their behaviour to that necessary to achieve comfort and security.

For herbivores, the supply of food and the cost of acquiring food is defined by the availability of pasture, for carnivores by the energy cost of hunting for a meal. For most adult animals, most of the time, food is in short supply so they need to conserve energy and do little more than that which is absolutely necessary to ensure their comfort and survival. The exception to this rule is, of course, sex when the motivation to mate overwhelms the desire for a quiet life.

There is however much more to the behaviour of animals with sentient minds than the basic need to obtain nutrition and ensure survival. The chicken that sleeps at night on a perch, out of the reach of rats and foxes, is motivated by the desire for security. This is simply a functional pattern of behaviour. However, the cat, or lamb, that basks in the sun, or the dog that lies so close to the fire that it gets too hot and starts to pant are clearly motivated by the desire for a strictly unnecessary experience that we may interpret as hedonistic pleasure.

This brings us to the question of play. Here again, the stern behaviourists argue that play, as observed in young carnivores, is a strictly functional pursuit. Cubs or kittens that spring out on each other and engage in mock fights may well be learning life skills but that does not exclude the possibility that they are also having fun. Other forms of play behaviour such as lambs or foals 'pronking' (bouncing on all four feet) or engaging in crazy races would appear to have no function other than the pursuit of pleasure. I was once asked to appear on a television programme to address the question sent in by a viewer 'why are cows so boring?' I explained that the dairy cow, the most overworked of mothers, has neither the time nor the energy for play. I then dropped a couple of balloons into a pen full of calves and at once initiated a game of nose ball. Play exists for sentient animals as a luxury pursuit open to those who have both the time for play and sufficient food to provide the energy for play. In former centuries, affluent English gentlemen, strongly advised not to go 'into trade' and having, in consequence, a surfeit of spare time, devised an impressive range of sporting activities. This is a classic example of the definition of play as a luxury pursuit in a sentient species.

Hope and Despair

We can accept that the responses of animals to primitive emotions such as fear and pain provide convincing evidence that they do not live only in the present, but can we extend this to include things that are normally considered expressions of higher human consciousness, namely hope and despair? Here again, I call on the authoritative voice of Wittgenstein (85).

'One can imagine an animal angry, frightened, unhappy, happy, startled. But hopeful, why not? Hope is an expression of belief, but belief is not thinking.'

If we interpret Wittgenstein's assertion about hope in the context of sentience then hope becomes a positive emotion, a feeling that through my actions or the actions of others, something good may ensue and I shall feel better. It may be an anthropomorphic error to view the concepts of hope and despair as we understand them to be the products of a conscious mind, aware that it is aware. Human development of the concept of hope may be much more complex than that of other animals and may extend, not always rationally, much further into the future, viz. '*the sure and certain hope of the resurrection to eternal life*' as expressed in the Anglican Book of Common Prayer. However, I have no problem in defining hope simply as an emotional feeling that the state of feeling good can be sustained or may, in future, improve. A simple illustration of the concept of hope in animals is provided by the primitive sensation of hunger. Calves that bleat, sows that chew the bars, or wild cats that pace their cages in anticipation of a meal are displaying an emotion linked to the expectation that a meal will arrive at the expected time. When the hoped-for event arrives, their behaviour is consistent with pleasure.

The concept that hope is a learned emotion, a feeling about the future based on past experience, is particularly relevant to domesticated animals since the pattern of their lives is so dependent on the actions of their owners. If a dog or farm animal is fed at increasingly irregular intervals or not at all, this confident expectation will be replaced by anxiety and this negative emotion will be greatest at former mealtimes. Similarly, the dog that is repeatedly disturbed by the non-return of its owner at the expected time may develop prolonged and incurable separation anxiety. The primate confined for years in a barren cage with little or no social contact will develop the signs (or non-signs) of profound apathy. These observations are reinforced by hard evidence from controlled trials with laboratory animals. Prolonged isolation or social challenges leading to defeat can inhibit neural development in mice leading to lasting changes in physical and behavioural development, such as decreased appetite, growth rate and signs of anxiety or apathy indicative of a profound deterioration in mood. The word to describe the emotional consequences of these failures of hope has to be despair.

Sex and Love

The function of reproduction is to preserve and multiply one's own genes, the most essential element of Darwinian fitness. However, the motivation to mate in animals is not driven by any long-term strategy for family planning but by the urge for sex. In many species, males are instinctively motivated to mate with any receptive female and this motivation can overwhelm even that to survive, most dramatically in species such as the red deer and the walrus that spread their genes most widely by maintaining and defending large harems. Females tend to be more circumspect. They are only motivated to mate at the time of oestrus and this motivation, of course, disappears when they become pregnant. In many species, the female in oestrus will actively seek out the male. In species like the red deer, males compete with each other to establish superiority and 'protect' a group of females in their own breeding ground or lek. However, they seldom compete for females in other leks. The females choose to enter and remain in a particular lek motivated mainly by their preference for a particular stag but also possibly to escape the attention of others (15). In these mammals, genetic fitness is promoted by competition between males. In altricial (nest fed) birds, where the young chicks are entirely dependent on their parent or parents, females tend to be selective: choosing males on the basis of their appearance or display. Here, it is the female that has most control over the process of genetic selection. Many species of altricial birds, where both parents contribute to the rearing of their offspring, form life-long attachments that have obvious survival value.

The hard-wired motivation to mate is an extremely strong expression of passion. However, we cannot be certain whether this expression of passion is associated with pleasure. Studies with rats have shown that the sex act stimulates an area of the brain defined (by us) as the pleasure centre, and rats with electrodes inserted at this site are motivated to elicit this response by self-stimulation (55). I am not convinced that we can call this pleasure, as we understand it, without resort to anthropomorphism. In some species, for example, cats, the shape of the penis appears to cause pain to the female during mating, but this does not affect her motivation.

The hard-wired sex drive in animals, simply defined by the motivation to copulate is a very primitive expression of the power of sentience. However, I have no objection to defining as love, the complex pattern of instinctive and thoughtful behaviour necessary for the successful rearing of the products of sex, whether by one or both parents. When both parents are necessary to the success of their offspring, as is the case for most birds, males and females constantly reinforce their bond through displays of mutual appreciation. This bonding will last, at least, for the duration of their responsibilities to one generation of offspring and, in some species, will carry on for life. However, as I shall describe in Chapter 7, staying together for the sake of the children does not necessarily equate to sexual fidelity.

Summary

The Buddhist five skandhas of sentience provide an elegant, biologically accurate system for classification of sentience within the animal kingdom. We can assume with confidence that humans possess all five but what about the others? How sentient are they and how much should we care? What about the amoeba? Because it reacts to stimuli likely to cause damage, does this mean it experiences sensation? It would be comfortable to avoid the issue altogether, an approach described, tongue in cheek, by Roger Brown (11). '*How much of the mentality that we offer one another ought we to allow the monkey, the sparrow, the goldfish, the ant? Hadn't we better reserve something for ourselves alone, perhaps consciousness? Most people are determined to hold the line against animals. Grant them the claim to make linguistic references and they will be putting in a case for minds and souls. The whole phyletic scale will come trouping into heaven demanding immortality for the tadpole and the hippopotamus. Better be firm now.'*

It should be clear by now that I am not prepared to be so dismissive as Roger Brown, although I do like his turn of phrase. My aim is to lay the foundation for our ethical approach to animals based on the best available evidence as to the nature of their minds. I believe sentience begins to matter when it includes the first two skandhas, matter and sensation, if only on the basis of the principle of reverence for life. The UK Scientific Procedures Act recognises that the property of sensation alone is sufficient to give protection to named species from harmful procedures conducted in the interests of science, health and safety. For species with sentient minds, that possess the powers of perception and mental formulation, our responsibility extends beyond the maxim 'do no harm' and should accept responsibility to respect their quality of life. This applies whether or not they demonstrate evidence of the fifth skandha, namely human consciousness. In subsequent chapters, I shall argue that many animals in many orders possess more than three skandhas, some scoring as high as 4.5. These properties of sentience have evolved in different ways to reflect the special needs of different species in their different environments. Moreover, we are discovering more and more evidence for these properties of deep sentience in the minds of all vertebrates and an increasing range of invertebrate species (21). Read on.

Special Senses and Their Interpretation

3

The tools that enable the animal to gather and interpret incoming sensations and information involve a number of receptors (eyes, ears etc.) and a processing centre within the central nervous system. The receptors with which we, as humans, are most familiar are touch, vision, hearing, smell and taste. These convey information about the physical and chemical environment. The eyes record light, the ears record sound, the nose and mouth record smell and taste. This information is carried to the brain where it does not enter an empty room but mingles with the mental images already in store. When a visual image, a face, for example, is registered in the cortex it will be recognised as familiar if that image is already logged in the memory bank or identified as novel by virtue of subtle differences from the other images available for inspection. The same principle applies to sounds and smells. This is a simple expression of Gestalt theory, which argues that animals do not build up mental pictures *de novo* but interpret images in a holistic way; something that can only be done if the mind has some prior knowledge of what to expect (23, 68). A good demonstration of this phenomenon is that humans who have had their sight restored having been blind from birth or shortly after, do not comprehend, at first, what they see. Gestalt theory gets a mixed reception from neurophysiologists but, like all theories, it makes no claims to be more than a partial explanation based on the available evidence. When we talk of minds, as distinct from brains, I believe it makes for a good working hypothesis. The amount of information transduced

Animal Welfare: Understanding Sentient Minds and Why it Matters, First Edition. John Webster.

by the special sense organs and sent to the brain is massive and would be overwhelming if the mind did not pay special attention to that which is important and reject the rest. Physiologists use the word *attention* to describe the ability of the mind to concentrate on what matters. In everyday speech, we are more likely to say we can 'focus our minds'. I shall use this expression from now on, while fully aware that it does not mean the same thing as to focus the eye. I believe Gestalt theory offers a good, parsimonious explanation of the mind's ability to prioritise the things that matter. The conscious mind can only give proper attention to the few details that are important if it is able to ignore the barrage of incoming information that does not matter at the time. I understand (from reading rather than experience) that some of the most alarming effects of psychotropic drugs such as LSD arise from the fact that this inhibitory process is impaired, resulting in brain overload where key signals are drowned out in the general noise.

The neurophysiological mechanisms involved in the reception and processing of incoming information by the special senses at the cellular level are essentially the same in all animals. However, there are large differences both within and between phyla (i.e. mammals, birds, fish, amphibians etc.) in the design and function of the special sense organs that pick up signals from the environment and the make-up of the central nervous system wherein the information is processed and decisions are made. Our human concept of the brain is that it lies within our skull and our large cerebral cortex is the key to our consciousness. This is not necessarily how other species interpret the world. Fish do not possess what human anatomists would recognise as the pain centre in the brain yet experience pain. The octopus has a relatively large brain yet perceives and interprets much incoming information using nerve cells in its limbs. Jellyfish have no organ that could be defined as a brain yet recognise light signals and use them to migrate between feeding grounds in the open sea and shelter in mangrove swamps. This reinforces my argument that a strict neuroanatomical approach to the understanding of the brain may fail to recognise many of the workings of the mind. The most direct way to discover whether an animal, can experience pain, fear, pleasure (etc.) is to ask the animal, using a non-verbal language that the animal can reasonably be expected to understand.

This chapter outlines the various ways in which animals identify and process information transduced by their special senses and how different species have, through evolution, given priority to the information most conducive to their own genetic success and downgraded or switched off others, not least to avoid information overload.

Vision

To open with a blinding flash of the obvious, the function of vision is to provide information at a distance in circumstances where and when there is enough light to see by. When we probe a little deeper, more profound questions emerge, such as: What is light? How much light? What sort of light? How does the brain triage the non-stop flow of information from the eyes to sift the small amount of information that matters from the general noise?

Light is defined as electromagnetic radiation within the visual spectrum. At first sight, this definition sounds clear cut but on closer inspection it becomes both subjective and species-specific. We humans define light in terms of the spectrum of electromagnetic radiation (EMR) that is sensed by the retina in our eyes, transduced into nervous impulses and interpreted in our brains in the form of images that we can describe as pictures. Birds can see EMR at high frequencies that we describe as ultraviolet since they are invisible to us. Humans sense low-frequency EMR in the infrared through their skin and interpret it simply as heat. The development of infrared cameras has enabled us to experience night vision by transforming heat radiation into a visual signal. Snakes have built-in infrared cameras that sense infrared radiation and process it into detailed information as to the size, position and direction of movement of potential prey. Frogs can catch passing flies with extreme skill by instant processing of visual information as to their direction and speed of movement but are apparently unable to recognise the fly when it is stationary. Both snake and frog have developed highly effective hunting skills on the basis of incoming EMR signals, but it is unlikely that either are seeing pictures.

There are two main types of photoreceptor in the retina, rods and cones. In most animals, the rods make up 95% or more of all receptors and are extremely sensitive to light (they can recognise a single photon). They are the major transducers of visual information, but they do not distinguish colour. This is the property of the cones, which are far less numerous (<5%) and require bright light to be fully effective. Humans and old-world primates are trichromatic; they have three types of cone, so chimpanzees see colour the way we do. Most mammals have only two types of cone and, in consequence, a more limited range of colour discrimination. Lest this be seen as evidence of higher development in primates, I would add that birds, and the reptiles from which they evolved, have four cone types, as do most insects, while some species of butterfly have five. The greater the number of cone types, the greater the power of colour discrimination. This is a nice example of the principle that species develop the senses and skills that matter most. Colour perception matters a lot to birds and insects that depend largely on vision in sexual selection; not only selection of the favoured male within their species but also the correct species with which to mate. The eyes of nocturnal and marine mammals that operate at low light levels are dominated by rods, so they have less colour vision. Owls that hunt at night have very large eyes and a high density of rods but also depend for their hunting skills on an exquisite sensitivity to sound.

Many cephalopods, e.g. the octopus, communicate emotions such as threat and sexual attraction, by changing colour. They recognise colour changes in another individual and respond accordingly, using cells sensitive to light and colour situated all over the skin that react and drive responses to colour changes without necessarily involving the brain. Seasonal patterns of breeding in birds are driven by hormonal responses to changes in photoperiod, sensed by photoreceptors located not in the retina of the eye, but within the brain itself (encephalic photoreceptors). These must sense changes in light transmitted through the bones of the skull.

The ability to focus the mind on signals that matter and ignore the rest is particularly important in the context of visual information since the potential 'noise' level is so great. An elegant example of this is seen in the foraging behaviour of chickens. Hens foraging under trees in the autumn among a thick blanket of leaves and other unimportant vegetable material have an uncanny ability to spot and accurately pick up seeds and insects. These are the only bits of visual information that their mind bothers to register. The ability to focus the mind on what matters is an essential property of sentience.

Hearing

The ability to hear enables animals to communicate through sound. Hearing is especially important to humans because so much of our communication is through the spoken word. Transmission of oral sound through the media of air and water is of obvious importance to communication in most other animals although inevitably restricted by limitations to their ability to create and understand the wide range of sounds required for speech. In general, the importance of communication by sound increases in inverse proportion to the amount of information that can be acquired through sight. Bats that hunt at night rely primarily (but not exclusively) on echo location to identify their prey. Owls fly silently so they can listen out for the sound of voles in the undergrowth.

The two most important properties of hearing are sensitivity to volume and pitch. The range in pitch from the lowest to highest note that can be detected by the ear is called the audio frequency range. In humans, the audio frequency range is about 20 to 18 000 Hz in youth but detection of the higher frequencies declines with age. Discussion of sound reception, in terms of frequency range, is pretty meaningless to most people; better to speak of pitch in musical terms. The musical equivalent of a doubling in frequency is one octave (ah, the beauty of physics). The audio frequency range for humans is therefore about nine octaves (there are seven octaves on a piano). Many animals can hear sounds at a frequency too high for human ears. For dogs, this is about one octave, for rats and mice about two, for bats three to four. Rats and mice communicate alarm signals at a high pitch that cannot be recognised by many predators, although, unfortunately for them, they can be detected by cats. Whales communicate at a distance through songs sung at a low pitch. Dolphins communicate at close range using high pitched clicks and whistles.

A good example of the ability of the human mind to focus the mind on what is important is the 'cocktail party' effect. We can, if we so wish, cut down the background babble and concentrate on what our neighbour is saying. The holy grail for manufacturers of hearing aids is to reproduce the cocktail party effect but they still have a long way to go because this discrimination is not achieved at the level of the receiver, the cochlea of the ear, but through central commands from the brain. Experiments with anaesthetised and conscious cats has demonstrated clear evidence for central inhibition of sensory signals in the auditory nerve of a cat (63). In one classic experiment, the cat was placed in a room where the only sound was the ticking of a clock. While the cat was at rest, each tick was recorded as an impulse in the auditory nerve. A mouse was introduced to the room and the cat was immediately on high alert. The sensory impulses up the auditory nerve stopped. The brain was not simply ignoring the sound of the ticking clock, it actually switched it off.

One further homely illustration of the ability of the animal mind to extract sounds of importance from the general babble: some years ago, my wife and I were watching television of an evening while our cat was, as usual, fast asleep on the sofa. It was a nature programme and again, as usual, animals were doing violent things to one another. The cat slept through everything until, from the set, came the alarm cry of a small bird. Our cat awoke at once, got up and went round to the back of the set to see where the sound was coming from.

Smell and Taste

Smell, as we understand it, is the sensation we get when inhaling volatile aromatic compounds present in the air. Taste depends on direct contact. The sensations of smell and taste are linked and, for most purposes, may be considered together under the general definition of flavour. In human experience, the special senses of smell and taste rank relatively low in importance. Sight and hearing are essential for our independent existence. Smell and taste can be vital to our survival from time to time (e.g. to sense a gas leak). At other times, they may be no more than a luxury (e.g. the taste of good food). In most other animals, the process of olfaction is much more sensitive, much more complex and much more important as a supplier of essential information. All sentient animals have the ability to detect and recognise smells (volatiles). However, sensitivity to odours varies widely between species. The noses of dogs, rats, pigs and bears, for example, are at least 10 000 times more sensitive and discriminatory than humans. Pigs and bears rely on their highly developed sense of smell to detect food sources hidden underground. Many (not all) breeds of dog hunt by scent and this ability has been enhanced by selection. The bloodhound appears to have the most sensitive nose of all. We humans have exploited these skills to good effect. For centuries, pigs have been used to hunt for truffles underground. Today we make regular use of dogs to sniff out drugs and explosives and are training them to diagnose disease conditions, such as cancers and now Covid 19. Rats are just as good as dogs and are proving valuable in mine detection, since they can stand and signal while over the mine without blowing themselves up. They would be equally adept at detecting diseases in hospitals but are, sadly, less patient friendly.

Grazing animals are also exquisitely sensitive to odours. In laboratory experiments, calves have been shown to detect the smell of salt (sodium chloride) at a concentration less than 10 molecules per litre (4). This facility reflects the fact that many grazing animals range lands far from the sea where the availability of salt, a vital nutrient, is severely limited. Rocky mountain goats transported to salt-deficient areas in Western Canada have been observed to identify and migrate further than ten miles to natural sources of rock salt in the mountains (Figure 3.1).

The importance and details of the information provided through the senses of smell and taste vary greatly between species. For many insects, olfaction is the principal source of information as to the location of prey, predators and mates. Fish use olfaction not only to identify other fish, conspecific predators and prey, but also for navigation purposes. Navigation in fish is considered in detail in Chapter 6. At this stage, it is enough to state that salmon use olfaction, together with other senses, not entirely understood, to return to the rivers where they were born. As they approach their destination, they are increasingly driven by the flavour of home.

A wide range of animals, including snakes, lizards and most mammals have a special mechanism for processing olfactory information. This involves a vomeronasal organ, sited between the nose and the mouth that gives them the added ability to 'taste' non-volatile compounds in the environment. Snakes pick up signals with their tongues, elephants sample the environment with the tips of their trunks, then touch their vomeronasal organ with their tongue or trunk to carry the information to the sensor designed to recognise it. The vomeronasal organ plays an important role in predator/prey recognition

Figure 3.1 Goats at salt licks in the Rocky Mountains. Raymond Gehman/Corbis Documentary/Getty Images.

but its most important and specific function is to sense and identify information carried by pheromones, best defined as organic compounds that act like hormones outside the body to transmit messages at a distance within a species. Pheromones carry alarm signals, food signals and sex signals, especially from females to indicate that they are ready to mate. Like hormones, which convey information at a distance within the body, they trigger hard-wired responses, not much influenced by reason. The vomeronasal organ is vestigial or absent in humans and old-world primates, which implies that, whatever we might imagine, we are not much influenced by pheromones.

Cutaneous Sensation, Touch

Cutaneous sensations are those arising from direct contact with the skin. These include touch, temperature, pressure, pain and itching. These sensations are identified and differentiated by different nerve endings. Sensitivity to touch, pressure, pain (etc.) varies according to the density of nerve endings in different regions of the skin. Our fingers, for example, are much more sensitive to touch than is the skin of our buttocks. In most mammals, the sensation of touch extends to the hair, especially specialised sensitive hairs such as the cat's whiskers.

At first glance, cutaneous sensation appears fairly straightforward. When we touch something with our eyes closed, we know it's there and we can get a good idea of its shape, consistency and temperature. In fact, the brain can do better than that. While idling away my days at Cambridge punting on the Cam, I was constantly struck by the fact that I could sense whether the bottom end of the punt pole, perhaps four metres away from my hand was striking gravel, rock or (mostly) mud. This is another example of the ability of the mind to build complex images from quite simple incoming signals on the basis of images already in store.

Magnetoreception

Magnetoreception describes the ability of a living creature to sense magnetic fields. We are not yet sure how. At present, the most plausible theory is that (especially) migrating animals like birds, fish, and sea turtles, possess a built-in compass, in the form of an accumulation of magnetite (magnetic iron ores). This property is widely distributed in life forms, from bacteria to mammals. In birds, the magnet is situated in the beak, in fish it appears to be distributed throughout the skin. A built-in magnet, would, in its simplest form, equip these animals with a compass that enabled them to determine direction. There is good experimental evidence to confirm that these animals are able to orientate themselves with respect to the earth's magnetic field, since they can be sent off course in a predictable way by artificially altering the field. Some suggest that magnetoreception, in its most advanced form, may involve a magnet that not only detects magnetic north from 'movement' of the compass in the horizontal plane but may also be able to detect latitude by interpreting lines of force in the earth's geomagnetic field in three dimensions (32, 82). This is a mind-boggling concept and there is a lot we have yet to learn. We have identified the property of magnetoreception in the species that need it the most: insects, birds, fish and sea turtles that migrate long distances, and subterranean mammals like the mole rat that cannot see where they are going. We are also pretty certain that we humans don't have it.

Beyond this point, it becomes really difficult. I can picture the geometry involved in the design of a three-dimensional compass that can determine both direction and latitude. To my knowledge, nobody has come up with a convincing physical explanation as to how a built-in magnet could provide even a rough indication of the north-south, east-west coordinates of global position that would indicate where a fish or bird was at every stage of its journey – yet they do seem to know where they are. The instinctive, hard-wired faculty of magnetoreception makes a key contribution to the navigational skills of migrating animals but it cannot be the whole story. The full range of skills required for short- and long-range navigation (of which magnetoreception can be but one) are discussed in later chapters.

Interpreting the Special Senses

Having outlined the receptors designed to identify and transmit sensory signals from the external environment, we must now consider how the minds of sentient animals interpret these signals and then decide what to do. All living forms from sunflowers to scientists react to stimuli and most of these reactions will be automatic or instinctive. Animals that possess the second skandha, the property of sensation, can recognise the signal as sight, sound, smell, touch or pain, and make an appropriate, programmed reflex response without necessarily recruiting the power of the sentient mind. The deep skandha of the sentient mind enable an animal to recognise the sensation as good, bad, or indifferent, choose a response from a range of options, and form a better idea of what to do the next time the sensation recurs.

We are now entering largely uncharted waters of the animal mind where it is impossible to avoid a lot of speculation. Some repetition is necessary here. To begin with, what we can say for sure: each sentient animal arrives with its mental birthright: a mind that contains both a data bank and a basic operation manual containing information necessary to operate as a newly arrived individual within the environment of its ancestors. In all animals, this contains the instructions necessary to operate primitive actions such as eating, drinking and locomotion. In animals with the inner circles of sentience that define the sentient mind, this includes a store of images linked to more complex and conscious reactions and responses.

As these animals grow and develop, whether or not under the protection of a parent or parents, they are presented with a continuous input of sensations and information that they must acknowledge, process, highlight as important or discard, thereby ever enlarging and enriching the picture of the world they inherited at birth. The more they learn from experiences, the more they develop their minds. Two of the innate senses available at birth are curiosity and caution. These motivate an animal to seek knowledge through experience while armoured with a suspicion of all that is new so not yet classified as good, bad or indifferent. The innate sense of neophobia is likely to remain throughout life but will be modulated, first, by how well the animal becomes accustomed to its environment and, second, by how well it learns to cope with challenges old and new. If the environment is reasonably stable and the animal learns that it can cope, then it should, most of the time, experience a sense of wellbeing. If it is presented with too many unexpected challenges and has difficulty in coping, then its mood is likely to shift out of the comfort zone towards anxiety or depression.

In Chapter 2, I made reference to 'The Master and his Emissary': the notion that when an animal is presented with a challenge or experience that calls for action, the right side of the brain forms an overall, emotional impression of the big picture then directs the left side to think about the details and come up with a sensible response. This is a memorable image, too simplistic for a neurobiologist with detailed knowledge of the location of functions within the brain but it does embrace three sound principles:

- the sentient mind is primarily driven to action by how it feels
- most conscious actions designed to maintain or restore a feeling of wellbeing involve some degree of cognition, or thought
- the feeling brain and the thinking brain are intimately linked: each modulates the response of the other.

The study of cognitive bias provides an excellent experimental approach to our understanding of interactions between the feeling and thinking brain. To give one example: rats reared in stable, enriched or barren, stressful environments were trained to associate sound cues with pleasant (food) or unpleasant (white noise) outcomes. The sound cues were 2 and 4 kHz (one octave) apart. Subsequently, the pitch of the sound cues was moved closer and closer to the mid-point where the rats were unable to tell whether the outcome would be good or bad. At this point, rats reared in happy environments were more inclined to expect good news, those conditioned by prior stress, more inclined to expect the worst (50). In simple terms, these rats had been conditioned by their experience to become optimists or pessimists. Similar studies have been carried out with starlings and dogs. When dogs with a history of separation anxiety are presented with ambiguous signals, they are more inclined to expect the worst. The study of cognitive bias provides clear evidence that mood can influence reason (51). This power of the feeling mind over the rational mind may well favour adaptation to the environment. Animals that have been chastened by past experience are more likely to select from the stacked shelves of sensory input those signals most likely to threaten their wellbeing.

We may safely conclude from the evidence that animals displaying the deeper levels of sentience learn from experience and make choices based on their emotional response to and rational understanding of their physical and social world. I believe that this can be fully expressed within the fourth circle of sentience, namely mental formulation, and does not necessarily require the possession of the inner circle, namely consciousness as we humans understand it. Many may dispute this conclusion. It does, however, have the practical merit that it enables us to avoid 'angels on a pinhead' type debates as to the exact nature of consciousness as unnecessary to the exploration of nearly all aspects of animal sentience in the context of their welfare and our approach to our fellow mortals.

Theory of Mind, or Metarepresentation

The inner skandha of sentience, usually equated to human consciousness, being aware that we are aware, enables us to recognise how we feel, our emotions, our desires, our fears and our knowledge. This recognition makes it possible for us to appreciate that other humans can experience the same set of feelings but are likely to interpret these feelings in their own way. Scientists refer to this property of consciousness as 'Theory of Mind' (21,40), a definition that, without further explanation, means nothing at all. Others use the term metarepresentation, defined as the ability to represent a mental representation (which I can just about grasp). If they were to use words that made sense to an intelligent layperson, they would simply say that it describes the ability of individuals to attribute mental states to themselves and, by extension, to others, i.e. 'I think I know what you are thinking' or 'I feel your pain, or joy'. Awareness of how others might feel is a powerfully adaptive tool in the business of social intercourse as it opens the possibility for expression of the higher social graces such as empathy, altruism and compassion. Equally, it is a prerequisite for deceit and fraud. At first sight, this may all seem rather obvious. However, children with autism have extreme difficulty in recognising the emotional state of others, and this is interpreted as an absence of the property of metarepresentation (1). It is argued from this that ToM is not a universal property of human consciousness.

If human consciousness does not always incorporate metarepresentation, what then of the other animals? What animals, if any, are able to behave in ways that indicate they can read the minds of others and predict the way they are likely to behave in social interactions? At present, the jury is out. There is some evidence that primates, dogs and corvid birds exhibit ToM, although many social interactions of these and other animals may simply be interpreted in terms of associative learning (40). They forecast the actions of others not by reading their minds but from observation and experience of their current and past behaviour. A crow does not need to know what another crow is thinking before deciding to cache its food when other crows are out of sight. A dog does not need to know what its owner is thinking before deciding it is unwise to steal food while its owner is in the room. These are rather simplistic examples, but more subtle experiments designed to test whether animals can make sound predictions as to, for example, the location of hidden food from the *unobservable* inner states of others have yet to prove entirely convincing. We need to exercise extreme caution before we attribute the property of ToM to non-human animals. Moreover, it may be unnecessary in order to experience a satisfactory social life. There are, however, some manifestations of behaviour, e.g. among the apes, that would be difficult to interpret any other way than as an expression of fellow feelings. Chimpanzees and bonobos will go out of their way to give comfort to a member of their group that has been injured or come off worse in a fight. This, which is described as affiliative behaviour, is a convincing demonstration of empathy in the form of compassion for a neighbour in distress.

Summary

All animals are born with a hard-drive operation manual, a set of sensory monitors and mechanical aids that equip them with a toolbox appropriate to the special needs and challenges of their life. The possession of a sentient mind enables an animal to interpret and enlarge this operation manual and make better use of these tools by learning from experience. This is achieved through a mixture of trial and error and knowledge acquired from observation of the experience of others. The need to learn from experience is especially important for species that live in an unpredictable, challenging environment, like the sewer rat, animals that have few natural defences, like the chicken, and animals that depend for their success on the establishment of stable social groups, like the chimpanzee and bonobo. I have now laid out the toolbox. The time has come to explore how sentient animals select tools from this box and develop their skills to meet their own special needs, where these are largely determined by the special challenges of their environment.

Survival Strategies

'It is not the most intellectual of the species that survives; it is not the strongest that survives; but the species that survives is the one that is able best to adapt and adjust to the changing environment in which it finds itself'

Leon Meggison

This piece of wisdom neatly encapsulates the Darwinian principle of survival of the fittest. Dinosaurs were extremely successful for 150 million years but failed to cope with a climatic catastrophe. The shark has managed very well without the need for change since the time of the dinosaurs because its environment hasn't changed. Homo sapiens who arose less than one million years ago, has been more or less dominant for only the last few thousand and, in the last 200 years, largely through our invention of machines running on fossil fuels, has put our families and other animals at risk by threatening the habitat for all forms of life. We may be more intellectual but, according to the above definition, we are not proving to be a great success. Adaptation to the environment requires possession and development of the necessary machinery (e.g. limbs, claws, beak), senses (e.g. vision, hearing), skills (e.g. hunting, hiding) and mental ability (cognitive understanding and emotional intelligence). Different species have developed different abilities according to their different needs, which means that the same animal can

Animal Welfare: Understanding Sentient Minds and Why it Matters, First Edition. John Webster.

be brilliant at some things, hopeless at others. The albatross appears to carry in its head a map of the southern hemisphere but fails to recognise its own chick if it falls out of the nest. It is meaningless to claim that one sentient species is more intelligent or more developed than another: each has developed the special senses and skills that it most needs. The corollary also applies. The wise saying, '*we are all of us ignorant, we are just ignorant about different things*' applies not only to humans but right across the animal kingdom.

All sentient animals are born with a foundation of emotional and cognitive intelligence, an inherited operation manual and toolkit of physical and mental resources designed to promote survival and encourage wellbeing. Animals with sentient minds are able to build on this birthright, develop knowledge and emotional understanding of their physical and social environment, and hone special skills. All these properties of deep sentience enable animals to develop strategies designed to favour personal survival and reproductive success but, I repeat, they also carry the potential to cause suffering if, for any reason, they are thwarted in their attempts to carry them out.

Success, in Darwinian terms, is measured by the extent to which a species, or population within a species, is able to promote the survival and spread of its genes. Translating this into the day-to-day business of life, it is measured by the ability of individuals to promote their own survival, reproduce and further ensure the survival of their offspring. This principle, of course, applies to all living things. Here, I consider, by way of a few examples, how animals employ the inner circles of sentience to polish their strategies for survival, reproductive success and social wellbeing. These include the ability to acquire food through foraging or hunting, the ability to cope with complex environments through innate and acquired skills in spatial awareness and navigation, and the ability to protect one's offspring until they are old enough to look after themselves.

Foraging

Charles Darwin is one of my heroes, not least because he considered all animals to be equally fascinating. In his later, housebound years, he wondered whether earthworms that pull fallen leaves down into the earth grab them at random or make decisions as to the most effective feeding strategy. He cut up paper into small triangles, observed the behaviour of the worms and discovered, to his great satisfaction, that they preferentially chose to grasp the paper at the point of an acute angle, thereby making it easier for them to get it underground. We may be confident this skill is hard-wired, each earthworm does not work it out for him/herself (earthworms are hermaphrodites). It is a skill and it is one that calls for a decision, but one that strictly follows the instructions in the ancestral operation manual.

Involvement of the inner circles of sentience in the development of foraging strategies can be found throughout the animal kingdom. Here, I select a few choice examples

Honeybees: There are over 16 000 species of bees, most of which are solitary: individual females build a nest, lay down a source of food consisting mostly of pollen, lay a clutch of eggs, seal the entrance and leave the eggs to hatch into larvae that eat the pollen and emerge the following year as mature bees. The honeybee, exploited by us for its ability to gather and store food for us at minimal cost to us, is an exception. It is described as eusocial, which means that the whole population works for the success of the colony. The structure and social organisation of a colony of honeybees is well documented, hard wired, so not relevant to my theme. However, their foraging strategies are relevant because they reveal special skills some of which are similar to other animals of the air, some uniquely their own. The foraging strategy of the worker bees operates according to the rules that apply for any animal that needs to forage for food, namely, to maximise the harvest, especially energy-rich food, relative to the energy cost of gathering that harvest. There are some variations in the way that bees go about gathering energy-rich nectar but, as a general rule, a number of scouts go out in search of a good source. Each scout returns to the hive and conveys her findings to the other workers in the form of a 'waggle dance'. This dance is performed on the vertical honeycombs inside the hide. It involves a series of figure-of-eight manoeuvres. She moves in a straight line waggling her abdomen from side to side, then loops back alternatively right then left to the starting point and repeats the sequence. The direction of her movement during the waggle sequence indicates the position of the food resource relative to the position of the sun (Figure 4.1). If the food source is directly in line with the sun, she dances vertically. When, as is usually the case, the food source is at an angle to the sun, she indicates the direction by adjusting her dance to the same angle from the vertical direction by dancing. The number of waggles in each sequence indicate the distance of the resource from the hive. The vigour with which she performs the waggles, and the number of repeats, give an indication of her estimate of the value of the resource (28).

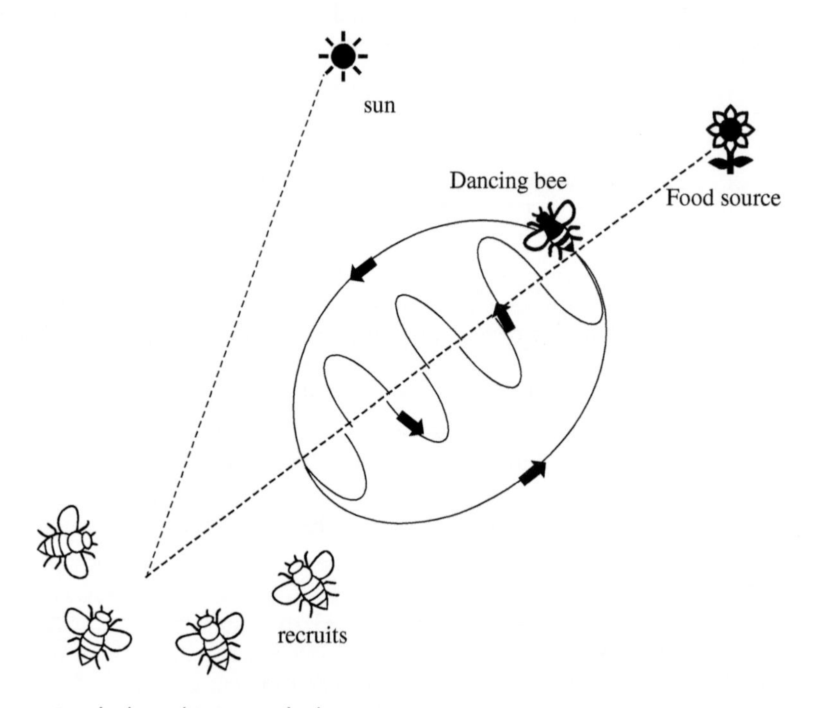

Figure 4.1 The honeybee's waggle dance

If all workers simply acted on information provided by a single scout, left the hive at once, got a bearing on the sun and followed the alignment to the sun indicated by the direction of the waggle sequence, they would all find the same food source at the distance indicated by the number of waggles. This is an impressive skill but not indicative of mental formulation. Moreover, it would not maximise foraging efficiency because that food source would quickly become exhausted. However, bee behaviour is not that robotic. The first variable is that the sun moves across the sky. Bees can take this into account because, like birds, they have a solar clock. A bee that sets out to forage in the afternoon will remember information as to direction relative to the sun conveyed in the morning and adjust for time of day. Moreover, the colony of individual bees in the colony do not respond *en masse* to a single piece of information. Several scouts will have been out and returned with information as to different food sources: their direction, distance and size of resource. Individual workers must then make decisions based on their estimate of the energy cost of travel to the resource relative to the likely reward measured by the size of the harvest. Different workers come to different conclusions, and this helps to ensure that the population achieves the most energy-efficient harvest of the entire locality. Clearly, individual bees make individual strategic decisions as to the costs and benefits of the foraging options on offer. In terms of information processing, the complexity of this degree of understanding merits classification within circle four; mental formulation. We must, I think, assume that this degree of understanding is latent within the birthright of honey bees. The fact that different individuals within the colony come to different decisions as to the cost-effectiveness of different foraging options is intriguing. It might suggest that they have the capacity to make judgements based on past experience. A more parsimonious explanation would be that different individuals are programmed to make different but consistent decisions based on different values given the proximity and size of the resource. This could be resolved by experiment.

Sheep: Given the opportunity to graze permanent pastures containing a variety of grasses, herbs and shrubs, sheep display an exquisite ability to select what is good for them. They avoid poisonous plants. This faculty, as with other animal species, is not simply based on how the food tastes when first they eat it. Shortly after eating a poisonous plant, they feel ill, remember this sense of malaise, associate it with the taste of the plant so learn to avoid it in future. They select nutritious plants, according to their specific needs for protein and energy, and there is some evidence that sheep with a high worm load will select plants with a high content of tannins having antiparasitic properties (41). They don't need a degree in nutrition to achieve all this; all they need is to be extremely sensitive to feelings that matter and have the capacity to relate these to some property of the food; initially taste but on subsequent occasions presumably a combination of sight and smell. The section of their operation manual that covers diet selection will be contained in their birthright. However, each sheep will make its own choices from the available menu of foods that is most likely to meet its nutritional needs and least likely to do harm. I shall have more to say on feeding choices in sheep in Chapter 8.

Tool use: When chimpanzees were first observed to fashion tools by stripping leaves from sticks to extract insects from a termite mound, it raised great excitement because it was put forward as a demonstration of the capacity of a human-like species to perform a behaviour previously assumed to be restricted to humans. Since then, once we started to look, we have found evidence that a wide range of animal species use tools when foraging for food that is not instantly accessible if, for example, it is hidden deep within a termite mound, or it has a protective coat, like nuts and shellfish, so needs to be got out of the tin.

Some birds and fish use stones to crack open nuts or break the shells of mussels. Wrasse, a marine fish, have been filmed opening scallops and clams held in their mouths by repeatedly swinging their heads to strike them against a nearby rock. Individuals probably learn this behaviour from observation of others, not necessarily their immediate family. The skills involved in tool use are far more advanced in corvid birds than primates. The most impressive performers in laboratory studies are the New Caledonian Crows. These birds display complex strategies for getting food from flasks, where it is seen but out of reach. These include fashioning a variety of complex tools, e.g. bending wire to make a hook, working out the correct sequence to solve an 8-stage puzzle, and getting at a piece of meat attached to a cork floating in water by dropping pebbles into the flask until the water level rises to the top of the jar (84). This skill was first described in Aesop's fable '*The Crow and the Pitcher*', which strongly suggests that his fable was not just a work of imagination. Aesop had either seen it or obtained information from a reliable source, which means that crows have been doing this sort of thing for more than 2 000 years while under no pressure from researchers. There is abundant evidence that crows in the wild devise a wide range of strategies to get at inaccessible food, like the meat inside molluscs. These include relatively primitive approaches like hitting a mussel with a stone or dropping it from a height on to a hard surface. My favourite is the behaviour of crows at popular seaside resorts who gather mussels, carry them to the car park, then carefully place them at a spot where they know they will get run over by a car tyre.

There is much scientific debate as to the complexity of the mental formulations necessary to carry out tasks where success requires a number of actions to be carried out in the correct sequence. Does the crow create in its mind a mental image of the whole sequence, or does it simply think one step at a time based on past experience of the things that can be done with the set of tools to hand? It would be interesting to know the answer to this question, but it is of little consequence to the crow. Crows in different areas develop different strategies to address high-priority needs, especially the need to maximise access to good food with least effort. This is clear evidence that these skills are learnt, rather than instinctive. Most crows pick up newly acquired skills from observation of others, but some undoubtedly have the ability to work out complex problems for themselves.

It would be a mistake however to assume that crows are universally smart. There are many reports of crows failing to solve problems that would appear as simple, or simpler to our eyes than fabricating hooks to get at inaccessible food sources. This evidence for limited, selective intelligence is consistent with my simple three-component model of the sentient mind, i.e. the genetically determined mental birthright and the modulation of this birthright by emotional and cognitive formulations acquired from experience. According to Gestalt theory, the operation manual with which the crow is equipped at birth includes a foraging strategy. Observations, such as the feeding behaviour of another crow, or the foraging potential of a new tool, or passing motor car, will register if they light up images already present in the original mindset. If they don't, and this applies, of course, to the vast majority of incoming signals, they will not register.

Apes and monkeys use simple tools like stones to crack nuts or strip the leaves from sticks to extract insects from termite mounds. However, there is little convincing evidence that primates *fashion* tools for specific purposes, so in this regard, they display less intelligence than the corvids. Here again, these skills are largely acquired by social learning and improve with practice. Youngsters learn how to use stones to crack open nuts from observation of adults. Different groups of chimpanzees develop different tools to extract insects from their nests. Some use sticks. Others make sponges from mosses. Individuals conform to group norms: they adopt the methods of their elders even when better tools may be to hand (27). Patterns of tool use in monkeys and the great apes are therefore an expression of cultural traditions. It is uncertain to what extent this social learning is a simple consequence of watching adults at work and how much can be attributed to education, i.e. formal instruction from parents and older siblings. I suspect there is a lot of proper education. In either event, it is a demonstration of deep sentience in the form of culture: adherence to the mores of the tribe, rather than an instinctive property of the species.

Storing the harvest: In habitats where there is a season of plenty and a season of shortage, animals need to adjust their foraging strategies to ensure survival during the hard months: winter in the temperate and boreal regions, the dry season in the tropics. The most primitive way to achieve this is to store nutrients inside the body, mostly as high-energy fat, during the season of plenty and conserve as much energy as possible during the period that food is scarce or unavailable. The survival strategy of the omnivorous bears of the Boreal forests is to eat and lay down as much fat as possible during the summer months then minimise energy demand by going into hibernation during the winter. Most of this behaviour is instinctive and mediated through hormonal responses to changes in day length. However, hibernation is not a passive process whereby hypothermia is a simple consequence of increased heat loss but a deliberate (if hard-wired) act of will. A signal from the hormone leptin (and possibly others) switches off the bear's voracious appetite. It retreats to the comfort of a winter den and then proceeds to slow down its metabolism. Both heart rate and metabolic heat production are reduced by about 15% *before body temperature starts to fall.* The physiological effect of this act of will is almost identical to that achieved by yogis embarking on transcendental meditation. It is common for bears to have 2–3 trial runs, or false starts, before true hibernation sets in, when heart rate falls from over 40 to under 10 beats per minutes and metabolic heat production falls to 25% of resting value in the summer months (19).

The pattern of bear reproduction has adapted to meet the seasonal changes in energy supply and demand. Female bears usually give birth every other year. Females without cubs mate in May or June but implantation of the fertilised ova is delayed until November when the female bear has retired to her winter den. Cubs are born in February, very immature and weighing less than 400 g. The mother, still half asleep, licks them into shape and feeds them with milk in the den for about three months until they emerge in the spring.

The alternative, more challenging approach to the challenge of avoiding death from starvation during the lean months is to build up a food store and ensure, so far as possible, that it is for personal use and not plundered by others. Perhaps the best exponents of this strategy are members of the squirrel family. One such member, the hoary marmot, or whistler, of Western Canada, feeds on a diet of grasses and leaves and hibernates for at least six months in colonies, where families can huddle together to conserve heat. They also build up food reserves by cutting grass and leaving it out to dry on rocks in the sun. While walking in the Rockies, when a newcomer to Canada, I was admonished by my knowledgeable companion for sitting down carelessly on a rock to eat my packed lunch and ruining a whistler's hay crop.

Red and grey tree squirrels collect nuts in the autumn and store them in caches. They are selective in their choice of nut, favouring, for example, varieties of acorns that store better because they are late to germinate in the spring. This appears to be an innate behaviour since it is observed in naïve individuals. However, squirrels also practice deception. They create dummy caches that have the appearance of a food store but are, in fact, empty. Moreover, they are more likely to manufacture dummy stores when population density is high, and the competition is greater. The capacity for deception is considered one of the most advanced expressions of mental formulation and may, indeed, indicate the property of metarepresentation, or Theory of Mind.

Hunting Behaviour: The Predator and the Prey

Hunting describes the process whereby an animal, or pack of animals, select another animal for their prey, approach it, kill and eat it. This requires a different set of skills from those employed in the indiscriminate actions of the ape that hauls insects out of a termite mound or the baleen whale that hoovers up krill from the northern waters. The motivation to hunt is obviously innate but hunting skills are learnt and improve with practice. Play fights between carnivore cubs provide early training in these skills but they are also fun. The adult, well-fed domesticated cat that plays with a captured mouse is also having fun. In the wild, the hunting of large herbivores by the big cats or packs of wolves is, in the strictest sense of the words, deadly serious. For the prey species, the threat is immediate: it must escape or die. For the carnivore, it must kill or starve.

Populations of wild herbivores have adapted to the threat of hunting. Indeed, the impala is less at risk of extinction than the cheetah. Nevertheless, we need to ask what might go through the mind of an individual impala or zebra in an environment where a pride of lions may emerge at any time. We shall never really know, but we can get some idea from their behaviour. All grazing herbivores, whether wild or domesticated, have a hard-wired programme within their minds that sets their *flight distance* from a potential predator. When the predator is outside their flight distance, they will be aware of its presence but see no need to take evasive action. As the predator approaches the flight distance they will, if unrestrained, move away in a controlled fashion in a direction determined by the angle of approach of the predator. If the predator encroaches within the flight distance, they will run off at speed. If they are prevented from escape by the walls of a pen or enclosure, they are likely to panic. A good way to appreciate this pattern of behaviour is to watch a sheepdog at work. When the dog is working well, the sheep move quietly in the desired direction. If it gets too close, or makes sudden moves, they will scatter. A good demonstration of the flight distance principle in operation in the wild is to observe (most probably on television) the behaviour of predator and prey at a water hole in the savannah. Zebras will drink from the same water hole at the same time as lions, but at what they perceive to be a safe distance. When the water hole shrinks after a prolonged drought, this becomes trickier. When forced to flee, individual behaviour is governed by two stark facts of life. The first is that, in order to survive, it is only necessary to run faster than other members of the group or to give the impression that you can run faster. The impala that leaps high in the air in the first seconds of escape is transmitting the message that I am a superior athlete and thus more likely to escape capture. The second fact of life is that you are only killed once. Survivors learn by experience that they can escape. Incipient threats may be picked up from a distance, possibly through the sense of smell, although lions will usually approach from down-wind. Most signals are likely to come from members of the herd who spot the danger first. The conspicuous white tail, or scut, on some species of deer would seem to be non-adaptive since it makes them more visible to a predator. However, the sudden flash of the scut of a deer disturbed by a predator conveys a message to the rest of the herd so favours the survival of the species.

It is, I think, fair to conclude from their behaviour that wild herbivores will be in a state of constant alert to the risk of letting predators get too close but will feel reasonably secure so long as they are able to maintain a safe distance. It may not be too fanciful to compare their state of mind with that of human pedestrians free-ranging the city streets. We are aware that motor vehicles are potential killers but feel secure so long as we stay on the pavement. Mothers, both human and herbivore, will be on special alert for their naïve offspring.

The hunting strategies of the predatory mammals are straightforward and involve little in the way of mental formulation beyond a degree of stealth in the approach. The big cats, like lions, and (especially) cheetahs, are sprinters with little stamina so make their kill quickly, or not at all. Wolves are stayers so run their prey to exhaustion. More elaborate demonstrations of subtlety and skill are seen in the 'lower' orders. The cephalopods (octopus, squid etc.) use a variety of hunting strategies designed to get them as close as possible to their prey (typically crabs) before making the final assault (75). The octopus expels ink and approaches under a smokescreen. Squid collect and deck themselves in coconut shells and use them as a disguise, both for defence and attack. More on this in Chapter 6.

Herons display a particularly advanced hunting strategy, namely the use of ground bait to attract fish. They have been observed to entice their prey with a wide range of ground bait, seeds, feathers, scraps of abandoned food. Individual birds pick up this skill from observation of others, not necessarily their parents. This strategy requires an advanced property of mind, namely the understanding required to work for an indirect or delayed reward. There is a particularly instructive YouTube video of children at the seaside throwing down bread for the birds. Most consumed the food on offer. One heron gathered bread, built up a small cache at the end of the jetty then cast it into the water where a group of fish had gathered. By denying itself the immediate reward it gathered a far better prize.

One of the most impressive displays of hunting skills is that of the archerfish (Figure 4.2). When it spies an insect sitting on a plant above the water, it spits a jet of water at the insect that knocks it down into the water and sets off at once to swim to

Figure 4.2 Archerfish strike. FLPA/Alamy Stock Photo.

the spot where it predicts it will fall. This exercise involves some complex geometry: first a correction for refraction of light at the water/air interface since the insect is not where it appears to be when viewed in a straight line from under water, second, the ability to calculate the height of the insect above the water and adjust the volume of its spit accordingly. While the key elements of this special skill will be hard-wired, these fish improve with practice. Moreover, there are personality differences between individuals. Some fish fire off at the first sight of an insect and their success rate may be below 50%. Others wait for the right moment, fire less often but achieve success rates in excess of 80%. A pedant may not classify spit as a tool, but this feeding behaviour must be considered as a demonstration of higher intelligence since different individuals adopt different strategies but all, through practice, improve upon the skills inherent in their birthright.

Spatial Awareness and Navigation

Any species that needs to forage in the search for food needs to know where it is going. Those that return to a nest or den to feed their offspring need to remember the way back. This requires a sense of spatial awareness and a degree of mental formulation necessary to create maps. The ability of animals to make maps was confirmed in a classic series of laboratory studies by Edward Tolman and his team in which rats had to find the correct exit from a maze in order to obtain a food reward (49). One group was given a food reward from the outset and learnt the correct path quite quickly. Initially, rats in the second group were given no reward at the exit and spent much longer wandering, apparently aimlessly, in the maze because there was no strong incentive to take a particular route, or indeed, to leave. The third group were given no reward and left to wander in the maze until day 11 when food was presented at the correct exit for the first time. These rats selected the correct path much more quickly than those in Group 1. Tolman called this latent learning; the ability to acquire potentially useful information from signals that carry no immediate reward. The rats had not just been wandering aimlessly but drawing a map of the maze, filed away in their mind to be recalled later for inspection.

These map-reading skills pale into insignificance when set against the natural map-making skills of birds. Young homing pigeons first learn their way about by following the lead of experienced adults. In successive training exercises, they are released further and further from home to enable them to draw larger and larger maps. If there are no colleagues to show them the way, they set their course for home from the position of the sun. If one is to navigate by the sun, one needs a built-in clock in order to tell the time of day. As any Boy Scout knows, you point the hour hand of your watch at the sun and south will be half-way between the hour hand and 12 o'clock (or one PM in summertime). In a beautiful series of experiments, scientists at Bristol University reset the body clocks of mature, well-travelled pigeons by confining them in a light-controlled environment. They were then taken out and released, wearing tracking collars so the scientists could follow their movements (7). These birds, confident they knew the way home, set off in the wrong direction. In what followed, they behaved exactly as we would. After a while they looked down, thought 'where am I? looked around for a conspicuous landmark, like a church spire, flew to this known position on the map, and reset a route for home. If they were 40° off course to the right, most then made a full course correction 80° to the left. This behaviour owed nothing to hard-wired instincts. The only explanation is that they had a map in their heads that showed them exactly where they were and from which they could plot the shortest route home.

Short-distance navigation is clearly based on recognised landmarks and mental maps. Migration, which is the most extreme demonstration of navigational skills, is essential to the fitness and survival of many species to ensure year-round access to their food supply and a secure and well-stocked breeding site. I shall discuss the migration of fish, birds and mammals to meet seasonal needs in Part 2, Adaptation to the Environment. Here, I consider only the tools that migrating animals may, or may not, recruit to achieve their purpose. Most current scientific opinion attributes the ability of salmon, for example, to migrate long distances to the innate property of magnetoreception (32, 82), reinforced, as they approach their breeding grounds, by the familiar smell of home. This hypothesis reflects the behaviour pattern of the conditioned scientist, which is never to stray beyond the limits of the evidence, but it is clearly not the full story. At present, this is beyond our imagination.

Much of my leisure time over the past 60 years has been spent at sea in small boats, mostly before I had access to reassuring information as to my coordinates (position), course and track provided by my global positioning system (GPS). For navigation, I had a chart to tell me where to go, a compass to tell me my course, (where I was pointing) and a pilot book to help me predict how the tides were causing my course to differ from my track (the direction in which I was actually travelling). I estimated my speed by dead reckoning, which, in my misguided youth, usually involved throwing an empty beer bottle over the bow and measuring the time it took to reach the stern. For most of recorded history, long-distance sailors aiming for a specific port on another continent relied on access to charts, a compass and a sextant from which to calculate latitude from a noon site of the sun. Precise calculation of longitude from the position of the sun and stars was only made possible by the development of chronometers that could keep perfect time.

Migrating birds are in a similar position to long distance sailors before the invention of modern navigational aids. They appear to possess a magnetic compass that can give them an indication of their course and possibly some indication of latitude from, respectively, the direction and strength of the magnetic fields (82). They also have a chronometer that gives them an accurate indication of time of day to enable them to steer by the sun (and possibly the stars). However, they also need to know the coordinates (latitude and longitude) of their destination and where they are at any point of the journey. For this, they need a chart. It requires too great a stretch of the imagination to assume that each swallow that migrates between Europe and Africa comes equipped with a detailed chart of the northern hemisphere. Moreover, even if it was able to mark on a chart the points of departure and arrival, this will not tell it where it is at any stage of the journey and what course adjustments may be necessary. They cannot calculate their current position by dead reckoning alone because it cannot account for the extent to which the wind has caused their track and distance travelled to differ from their set course and estimated speed through the air. Migrating fish are similarly affected by tides and currents. However, somehow, they manage. Migrating birds, blown seriously off course, or deliberately moved hundreds of kilometres from their flight path, can reset their bearings and successfully reach their planned destination.

Birds that migrate mainly over land tend to migrate in flocks: those on their first flight flying with the experienced adults, who probably have the memory of conspicuous sights *en route* such as the presence and shapes of mountains and coastlines. When it comes to animals like fish, turtles, birds, seals and whales that migrate long distances successfully across the featureless seas, my mind boggles; and I am not alone. It is difficult to avoid the conclusion that migrating animals not only know the coordinates (latitude and longitude) of their planned destination but can plot their coordinates with sufficient accuracy at any stage of their travels. This implies that they also carry a GPS, based on special senses that we humans do not possess and can scarcely imagine.

Breeding Behaviour and Parental Care

While sex, the act of copulation, may largely be a mindless process, success in child-care is likely to involve a high number of skills. In all species, most breeding behaviour, both sexual and parental, is instinctive. In most mammalian species, the old cliché, '*woman is monogamous, man is polygamous*', has a sound base in genetics. The number of offspring a female can rear is limited by the number of ova she can produce. Moreover, the energy demands of rearing each offspring to the point of independence is high. Assessed simply on the basis of the number of spermatozoa, the capacity of the male to pass on his genes is almost limitless. When the male plays little or no part in child rearing, he can expend most of his energy in maximising the size of his harem and keeping away other males. This, of course, can be hard and dangerous work. Stags at the end of rut are often emaciated, injured and more likely than hinds to succumb to the stresses of the winter.

In altricial birds that feed their young in the nest, breeding success depends on hard work from both parents from the time they start to build a nest until some time after fledging. There are exceptions, like the cuckoo, but they need not concern us here. The daily energy expenditure of passerine birds during the nest-feeding period exceeds that of any other homeotherm. When scaled according to body size, the only species that come close are long-distance racing cyclists during the Tour de France and the overworked dairy cow (77). Genetic fitness demands that male and female stay together and work together throughout the breeding season. The importance of this bond increases in proportion to the number of eggs laid and the amount of time and energy necessary to rear each chick. The most extreme demonstration of this is seen in the albatross. A typical pair will rear one chick every three years and, in the year that chick is born, devote over 200 days, first incubating, then feeding that single chick in the nest. The genetic fitness of the albatross owes little to the urge to mate and almost everything to parental care.

The hard-wired drive of the genetic imperative in altricial birds, which account for more than 90% of all species, requires both parents to be devoted to their offspring, and thereby each other, at least for one year and in some cases, e.g. the albatross, for a lifetime. The breeding behaviour of birds takes two forms: courtship, where the flamboyant male advertises his wares and the more sensibly attired female makes her choice; and bonding behaviour, where established couples reinforce their relationship through mutual displays of affection. Before this section gets too sentimental, I should point out that whereas over 90% of altricial birds practise social monogamy, up to 30% may not be the genetic descendant of the male that supported them. Some authors suggest that the proportion of altricial birds that practise strict sexual monogamy may be as low as 10% (81). Sexual morality is not critical to their breeding success; it matters only that the chicks are raised by two responsible adults.

Precocial birds are those mature enough to feed themselves from the moment of hatching although, in nature, they will still depend on their mother hens for warmth, security and education. The number of species is small and restricted to those adapted to life spent mostly on the ground. It is no surprise that the species we have most successfully farmed for food are precocial, ground-living birds like chickens and turkeys. They are easier to rear and easier to catch. Because the rearing costs to the parent are much less than for altricial birds, chicken cocks and turkey stags can afford to support harems and the hens have adapted to the social demands of living in groups.

All the patterns of reproductive behaviour described so far may well be hard-wired; conforming strictly to the instruction manual designed to promote genetic fitness. However, the motivation of individual birds, and couples, to carry out any action will be driven by an emotional need. Courtship behaviour by the male, at least, is a highly emotional business. The emotional response of the female to this display: her decision to choose the one who impresses her most by way of his appearance and/or behaviour, is clearly associated with breeding success. The behavioural aspects of courtship behaviour are probably more important in species that share the parental responsibilities.

The behaviour of parents in defence of their nest is highly charged with emotion. It ranges from furious and heroic attacks on predators much stronger and more dangerous than themselves to equally heroic acts of distraction behaviour, classically demonstrated in the broken-wing display of birds as they simulate injury and draw the predator away from the nest. An alternative form of deceit is for the parent, at first sight of the predator, to leave the nest and settle down to brood at a discrete distance then fly off as the predator makes its assault on nothing at all. Distraction behaviour is most often seen, rather obviously, in ground-nesting birds like plovers and waders, but other species have been seen to leave their nests in the trees to distract potential predators by feigning injury on the ground. This behaviour shows that these birds have the advanced mental capacity to practice deceit.

Breeding success, particularly success in chick rearing, undoubtedly has a cognitive element. There is plenty of evidence for this, but my favourite illustration comes from personal experience. At the dairy farm where I worked in the year before going to university, there was a bantam hen who, I was told, had previously lost two clutches of chicks, killed in their first days of life, by Gyp, the farm lurcher. While I was there, she brooded and hatched out a third clutch deep between hay bales in the Dutch barn and was only seen out with her chicks under close supervision when they were about three weeks of age. These chicks all survived. I don't know how they fed in the early days. Maybe she took them out at night on brief trips to forage for corn. There was plenty lying about. How she managed is not the issue. The fact is, that acting on the experience of her earlier failures, she modified her strategy and she succeeded.

5 Social Strategies

In the natural world, the principle of the survival of the fittest operates on the basis that each species, each family within that species, and each individual within that family acts instinctively to promote its genetic inheritance. While this is a great truth, it is not the whole truth. Animals with sentient minds are not just driven by their instincts, they also develop learned strategies designed to promote the wellbeing of themselves, their families and, in some cases, their social groups. Individuals of all species, if they are to thrive, will always need to compete but many will also need to collaborate if they are to achieve this sense of security and wellbeing. They need a strategy for social living, and this will be determined by the degree to which others (both other species and conspecifics) may favour or threaten these aims. Among the large carnivores and other killers, such as lions and bears, most individuals view other animals either as meat, competitors in the pursuit of meat, or a threat to their own genes. Lionesses and she-bears devote their lives to protecting their young from threats, the greatest of which can come from males of their own species seeking to promote their own genes by killing cubs sired by another male and mating with their mother. Smaller carnivores that hunt in packs establish extended family groups. A typical wolf pack is made up of a dominant breeding pair, the alpha male and even more dominant female with their extended family of sons and daughters who respect the authority of the alpha pair and look after each

Animal Welfare: Understanding Sentient Minds and Why it Matters, First Edition. John Webster.

other's interests. Here again, this is driven primarily by the genetic imperative but developed by a process of learning and education. This sense of deep family loyalty has been retained by the domesticated dog, although often misunderstood or abused by us. More on this in Chapter 10.

The survival of prey species, whether on land or sea, is often favoured by living in large groups. This is particularly important for animals that live out in the open, whether on the African Savannah or in the ocean. At the most basic level, group living need not involve any element of cooperation. If an animal under threat from a predator is unable to hide, the odds of getting eaten are greater if it is isolated than just one of a mob. The instinctive behaviour of many fish species under threat from sharks or dolphins is to swarm. It is likely that swarming behaviour in fish is entirely instinctive and operates according to simple rules governing the position and movement of one fish with respect to another. Any higher element of sentience would see all fish struggling to get into the middle of the mob. Swarming can lead to a feeding frenzy wherein many thousands may be killed but, for the individual, the odds that its genes will survive are improved. Once again, the sublime hindsight of Darwinian theory is seen in action.

Perhaps the most dramatic and beautiful demonstration of swarming behaviour is seen in the murmuration of starlings (Figure 5.1). Thousands of birds gather at dusk and create spectacular moving pictures in the sky before settling down together for the night. There is no evidence that this ballet is choreographed by a leader or leaders: each starling moves according to a strict set of rules regarding its position relative to 6–8 of its closest neighbours. There have been reports of multiple deaths in starlings when the leading birds have pulled out of a dive too close to the ground causing followers to crash. I know of no satisfactory explanation for this behaviour. Some have argued that, like fish, they instinctively swarm to reduce the individual risk of predation from raptor

Figure 5.1 A murmuration of starlings. Steve Littlewood/Photodisc/Getty Images.

birds. This fails to explain the complex rituals that they practise at dusk, especially in the winter months. A convincing, but unproven, suggestion is that the birds congregate before settling down to roost for the night in order to conserve warmth.

There is no direct evidence to support the suggestion that starling murmuration may be an expression of deep sentience in the form of affiliative social behaviour, or companionship. However, there is good evidence for affiliative behaviour in another flocking species, the rook. Rooks are highly social birds. The collective noun for a gathering of rooks is a parliament, which can amount to as many as 60000 birds. Within this enormous group, pairs establish close bonds. They mate for life and regularly reinforce their bonds with mutual displays of affection. However, they also establish close links with neighbours outwith immediate family, cooperate to defend the neighbourhood from outside aggression and display empathy in the form of compassion (affiliative behaviour including a form of kissing) to members of the family or neighbours injured or in distress (13). This is one of many demonstrations of emotional sentience of birds in the corvid family that are at least as advanced as those seen in primates. In my illustration of the five skandhas of sentience (*see* Figure 2.1), I suggested that this behaviour may indicate possession of the inner circle of sentience, namely (human) consciousness.

Vulnerable species of the open plains, like sheep and wildebeest, clearly favour individual survival by living in groups. Once again, most of this behaviour is likely to be instinctive: those of their ancestors that chose to live in company were more likely to survive so this has become the behaviour that works. This herding behaviour does not require these animals to establish any emotional bonds outside the family, it only requires them not to be seen as a threat. It does however require individuals within the herd or flock to learn to live together in order to ensure a quiet life. Problems of living together are compounded by domestication, particularly the domestication of farm animals, where large numbers of pigs, sheep and cattle, and enormous numbers of poultry are compelled to high-density living, with little or no opportunity to create their own personal space.

Sentient Social Life

Expression of the inner circles of sentience in the social interactions of animals were considered in Chapter 2. Here, I review the strategies adopted by social animals to meet their needs for biological success and a sense of emotional wellbeing. We can make a distinction between species that simply congregate in large numbers for instinctive reasons that favour survival, and those that appear to require the presence of a sentient mind. Quality of life within a flock, herd or tribe of social animals will depend on some or all of the following:

- Stable, non-harmful social relationships
- Communication of emotions and information
- Social learning acquired through observation and formal education
- Emotionally advanced expressions of social behaviour such as friendship, empathy and compassion

Social Hierarchies: The Pecking Order

The first essential for a satisfactory social life is to learn to get on with one another. The expression 'pecking order' is in common use to describe the establishment of a hierarchal structure in animals, including human children and adults compelled to live together in environments such as boarding schools, offices or the armed forces. In regimented societies such as the armed forces, a stable pecking order is established by strict attention to rituals imposed by an established hierarchy. In schools where the hierarchical structure is less well defined there is a greater tendency towards anarchy. The phrase 'pecking order' derives, obviously, from chickens, where the social structure of the population is hierarchical and can involve aggression. Different chickens express their personalities in different ways, and these personalities develop as they work out how they should interact with other individuals within the group.

The sustained use of direct, injurious conflict to maintain the social order is, by definition, non-adaptive for all parties. In animal societies, most encounters involve non-injurious ritual displays where both dominant and submissive individuals demonstrate that they know their place. Unless sex is involved, cattle and most social herbivores that live in large groups establish stable hierarchies with little aggression. Pecking orders in pigs (which, in the wild, form small social groups) tend to be unstable and this can create problems in commercial units.

The decision to submit to an opponent after the briefest of ritual interactions can be a good strategy for a quiet life. The hyperactive aggressive behaviour of some animals near the top of the pecking order would appear to stress them more than those lower down who have chosen, without fuss, to submit. In most circumstances, submissive individuals are able to adapt their behaviour in order to ensure they get what they need, like food. This was demonstrated by Keith Kendrick in a classic experiment where two pigs – one dominant, one submissive – were put together in a pen where they had to press a panel in order to receive a fixed amount of food. When the panel was close to the food trough, the dominant pig pressed the panel and immediately laid claim to the food. The submissive pig held back until the dominant pig decided that it had had enough. The press panel was then moved to the other side of the pen. The dominant pig continued to do all the pressing but now had to rush back and forth between panel and trough. The submissive pig stood close to the trough, moved in to take a few mouthfuls while the dominant pig was rushing off to press the button again then stepped quietly back when it returned.

While most dominance/submission interactions are resolved without coming to blows, they can result in injury. Aggressive feather pecking by dominant chickens and turkeys can, in some cases, be lethal. In small flocks of less than one dozen birds, given plenty of space and an enriched environment, this is seldom a problem. The occasional rogue aggressor can be removed if necessary. However, aggressive, injurious feather pecking can be a problem in commercial flocks of laying hens, and this has become a bigger worry as, in the pursuit of better animal welfare, more and more hens have been taken out of cages and allowed to range freely in large groups. There are many reasons why chickens engage in feather pecking and they do not necessarily involve aggression. More on this in Chapter 10.

Communication

For a group of animals to succeed as a family, flock or tribe, they must learn to communicate both information and emotion with each other. Without this, they would become a disorganised mob. Humans possess an impressive array of internal and external tools to facilitate communication, including speech, the written word and audio-visual aids. We also rely, to a great extent, on body language: not just extreme gestures like the arm waving of television presenters flailing away to generate visual interest, but subtle signals indicative of emotional state, such as small changes in facial expression.

Human speech conveys information by creating a near infinite variety of different sounds that we (if we have learnt the language) are able to interpret. This is undeniably intelligent. Emotion is conveyed by the quality of the sound. In speech, this adds to the message through the tone of voice: music, whether vocal or instrumental can be interpreted as pure emotion. A popular way to study intelligence in animals is to observe their capacity to recognise and understand human language. Dogs can recognise and respond to human spoken words and commands. The most charitable estimates suggest the number is into double figures, but this is hardly impressive by human standards. Moreover, so far as the dog is concerned, most of the information it needs will come not from the words themselves but from the emotional message conveyed by the tone of voice and body language of the speaker ('Gooood boy' in baby talk). The most impressive demonstrator of the ability of an animal to speak and understand human language has been Alex, the grey parrot studied by Irene Pepperberg, who could speak and understand more than 100 words and had an understanding of concepts such as bigger and smaller, above and below (57). Alex also asked questions. While the ability of dogs to communicate with humans may be impressive and that of Alex truly amazing, studies of this sort are not strictly relevant to the properties of mind that matter most in the context of animals that communicate entirely satisfactorily without recourse to speech.

Birds create an astonishing variety of sounds to convey a small number of simple messages like this territory is mine or I want sex. While the fundamentals of bird song are innate to the minds and vocal apparatus of the species, birds develop their vocal skills by listening to others. The proof of this is that they develop regional accents (43). Vocal mimicry is common, although some birds are better mimics than others. Captive parrots and mynah birds mimic a wide range of non-bird-like sounds, words, telephones, and even chainsaws. Wild starlings mimic a wide range of sounds, mostly other bird calls, often with a twiddly bit at the end to express their own personalities. There is a good biological reason why songbirds should develop their skills, because it increases their chances of attracting a mate. Some nesting birds may issue non-bird like sounds to confuse potential predators. However, many bird calls are difficult to accommodate within the mind of the stern scientist who believes that all animal behaviour must have a functional significance. It may just be that they are expressing pleasure, or, in the case of the caged bird, just doing something to pass the time on another boring day. As a young vet, I would frequently be called to carry out a post-mortem on an animal that had died and been taken to the knacker yard. Having finished, I was often invited into the house for a coffee or something stronger with the manager, a profoundly humane individual who treated every one of the sick and injured animals brought to his yard with the greatest of kindness and respect. He also kept an eclectic selection of pet animals, including several large dogs (he had a lot of free meat), a cat and a mynah bird. If the bird saw the cat outside the window, it would call 'Come on in pussy. The window's open!' – which it wasn't, so provoked a killer stare from the cat outside. This bird had obviously been trained to make this call, but it seems that he only did it when he saw the cat. He must also have observed that it got a response from both the cat and us. The cat scowled and we laughed. This implies that this bird was also communicating with cat and us in a way designed to alter our state of mind. This suggests metarepresentation.

Mammals communicate most information that matters through a combination of sound and visual signals, much of which carry a specificity that we would not recognise. Oft times when gazing with interest on a flock of sheep, I have observed a lamb bleat, one ewe at a distance look up, respond with a reassuring baa, and the lamb then run to join its mother. Dolphins communicate with each other using their own call signatures. If they want to contact a particular colleague, they will call it by name (64). Most mammals communicate a wide range of emotions through subtle facial expressions. Some, such as the dog or horse that flattens its ears to indicate submission, are apparent to us. Other, more subtle changes in facial expression have been identified by diligent scientists, mostly graduate students, after watching videos for many hours. For animals of the same species, their message will be immediate and clear.

Chimpanzees, the most studied of the great apes, have many effective ways of communicating information within the group and emotions relating to things that matter (41). Signals and information are conveyed by sound, gesture and facial expression in ways that make perfect sense. Sound signals are used to convey information at a distance, e.g. the threat of a predator or the location of a new food source. Gestures and body language, are used to encourage social intercourse. Several gestures carry an invitation to grooming. Leaf chewing signifies an invitation to sex. Facial expressions are used to convey emotion to others at a close distance. Bared teeth signal the threat of aggression. Grinning can be interpreted as submissive, ingratiating or simply being friendly, i.e. it conveys the same message as a human smile. These signals may be simple, but they are consistent and specific, so they are not ambiguous. One of our major problems in communicating with animals is that we talk too much. More on this later.

Cooperation and Empathy

If social animals are to thrive, they must cooperate. Cooperation between animals for whom individual success depends on the success of the population is seen throughout the animal kingdom. In its most basic form, cooperation may be motivated entirely by self-interest or the genetic imperative to promote the interests of the immediate family. Cooperation in colonies of ants is achieved through a simple series of fixed responses to specific signals. At this level, cooperation is assumed to be entirely instinctive. However, the evidence that honey-bees make choices as to which scout to follow to a food source suggests that this might not be the whole story. Flexible cooperation, fine-tuned to the circumstances of the moment, requires some degree of awareness of the feelings, thoughts and plans of others. As described earlier, this may be communicated directly from observing their expressions and behaviour and interpreting these as indicators of their intention and emotional state. If this were all, it would amount to no more than an intelligent interpretation of social signals acting in the cause of self-interest. It is, however, important to consider the possibility that cooperation in some sentient animals involves an element of empathy, expressed in the form of affiliative behaviour, i.e. consideration for others. The first requirement for true empathy is a recognition of self and non-self: 'theory of mind' or metarepresentation (21), which makes it possible for an animal to realise that, just as my behaviour is dictated by my own mind, your behaviour will be dictated by yours, and we may not think and feel the same.

The classic test to determine whether an animal can be aware of itself as an individual is the mirror test (22). An animal is accustomed to looking in the mirror (but at what?). A cross is painted on its forehead. If the next time the animal looks in the mirror, it notices the difference and tries to remove the mark, it is said to have passed the mirror test. It has recognised the image as an altered image of itself and this is cited as evidence of the ability to grasp the concept of self and non-self. Great apes pass this test. When they look at their image in a mirror, they recognise it as themselves and, probably, give most attention to the face. This ability to recognise self and non-self is shared by a few other animals, for example, elephants. However, the ability to identify someone else as someone else falls some way short of the ability to enter into someone else's mind and conclude 'I think I know what you are thinking or how you feel'. Many experiments designed to test the theory of mind are unable to exclude the possibility of associative learning i.e., 'I predict what you will do next, not because I can read your mind but because you have conveyed by a series of more or less subtle signals a message that I have learnt by experience to precede a particular action'. A more convincing indication of theory of mind is evidence of affiliative behaviour: unselfish concern for others. When a death occurs within the group, elephants, apes and many other primates, display behaviour that looks to us like grief and mourning. Moreover, this behaviour can extend to the whole group, not just the immediate family. It is a clear expression of emotional distress at the loss of a member of the group, but we cannot entirely dismiss the possibility that it is no more than an expression of the primitive emotion of fear: i.e. the death of this other individual may signify an increased threat to me. More convincing evidence of true empathy in primates is provided by the demonstration of compassion, most beautifully expressed in the form of consolation behaviour towards victims of aggression, especially when there has been a failure of reconciliation between the two warring parties. There is firm evidence for affiliative behaviour (consoling a colleague in distress) in the primates (chimpanzees and bonobos, (67), some evidence for marine mammals (dolphins), and a strong probability that it occurs in rooks. It would be difficult to interpret this behaviour in terms other than 'I know how you feel, and I care.'

Social Learning, Education and Culture

Social learning is a major contributor to the development of the sentient mind, especially in vulnerable species for whom the trial-and-error approach carries a high degree of danger. It is no coincidence that most social species are vulnerable species. Naïve animals acquire the majority of their survival skills from observation and mimicry. Tits learn from others how to remove the tops from milk bottles. Herons learn from others, not necessarily their parents, how to use ground bait to attract fish. Chickens spend more time observing the behaviour of dominant birds, presumably because it makes more sense to copy the habits of the winners. Naïve animals within populations of long-distance travellers, birds, wildebeest, elephants, follow those who know the way until such time as they have drawn their own maps. All this social learning creates in populations of birds, as in the great apes, a culture wherein the behaviour of individuals conforms to the mores of the society.

There is good evidence that social learning can involve more than passive observation and mimicry: it can be enriched by the formal business of education. Much of this evidence comes from studies with birds. In some birds, such as the Southern Pied Babbler, the parents prepare their chicks for fledging by training them to associate the arrival of food with a distinct call. Then, when they deem their offspring old enough, the parents entice them to leave the nest by making the same call from a distance (60). Other birds start to condition their chicks to respond to familiar sounds, such as the arrival of parent with food, before they even hatch.

While these exercises in formal education make a critical contribution to the mental development of the individual chicks, they are hard-wired into the behaviour of the parent. However, birds can do better than this. They can pass on to their offspring knowledge and understanding acquired in their own lifetime. In one of a series of classic experiments done by my colleague Christine Nicol, she induced a food aversion in bantam hens by linking the inclusion of lithium chloride with a particular food dye, yellow or blue. She then let the hens rear a clutch of chicks and turned hen and chicks out to forage on grain dyed yellow or blue. At this stage of the experiment, neither colour of grain had been dosed with lithium chloride, so both were innocuous, *but the hens didn't know that.* In consequence, they did everything in their power to discourage the chicks from eating the food that they had learnt from their own experience, would do them harm. This involved walking ahead of the chicks, only pecking at the 'safe' coloured food and actively discouraging them from approaching the other (54).

Social behaviour and social learning within a population, through observation, mimicry and education, promotes the development of standards of conduct and patterns of behaviour that are features of the group, rather than innate features of the species. These may fairly be described as mores and culture. We don't know how much of this may depend on simple observation and mimicry, and how much can be attributed to education, i.e. formal instruction from parents and older siblings. I suspect there is much education. In either case, it is a demonstration of culture: adherence to the mores of the tribe, rather than a property of the species.

Territorial Behaviour and Tribalism

Cooperation and mutual understanding within the group carry obvious benefits for social animals. In large groups of herbivores, like sheep and wildebeest, where the serious threats come from predation by other species, the larger the group the better. When different groups of the same species inhabit the same environment, there will be competition for resources and sexual partners. Groups, or individual mammals, typically establish their territories by scent marking the boundaries. The large cats and bears rub themselves against trees. Wolves urinate on posts. Farley Mowat gathered evidence for his splendid book. 'Never Cry Wolf' by camping in close proximity to a wolf pack in the Canadian forest (53). He established his territory by urinating every evening on trees and shrubs along the perimeter of an area round his tent that he designated for himself. When wolves approached his territory, they would pick up his scent and follow his directions to circumvent it. Conflict arises when outsiders fail to spot or choose to ignore these territorial signals.

The design and behaviour of the primates that share the tree canopy of the tropical forests illustrate a variety of strategies for resolution of competition for resources. Different species with different designs and different physical skills have adapted to different environmental niches. Chimpanzees live and feed mostly within the canopy, the heavy vegetarian gorillas more often at ground level. This does not, however, resolve the problem of competition for resources within species living in close proximity and this frequently creates the potential for conflict. In a social species like the chimpanzee, this threat leads to behaviour than can best be described as tribalism. Close bonds that extend beyond those to the individual family to the whole group, or tribe are accompanied by fierce opposition to neighbouring tribes. The nearest neighbours arouse the strongest feelings because their proximity makes them the greatest threat. Such tribalism may be considered an advanced form of sentient behaviour in non-human species, but primitive by human standards. However, we should not overestimate ourselves. Think only Arsenal vs Tottenham or Celtic vs Rangers. G.K. Chesterton reminded us that the Bible commands us to love our enemies and love our neighbours as ourselves. He added that this was not too difficult to achieve since they were usually the same.

Part 2

Shaping Sentient Minds: Adaptation to the Environment

6 Animals of the Waters

In Part 1, I explored the properties of the sentient mind, the special skills and the strategies employed by animals to promote not only survival and genetic fitness but also a sense of wellbeing and quality of life. Here, in Part 2, Adaptation to the Environment, I examine how sentient minds have been shaped by the special circumstances of their environments. I begin this voyage of exploration in the waters, partly because it is from the waters that so much life has emerged and partly because life in the sea, and on the seabed, has been relatively unaffected by human interference. Unscrupulous fishing practices have caused great damage to population numbers and pollution great damage to habitat. However, when we are not seeking to kill them, we leave them alone to get on with their lives (farmed fish are, of course, an exception). Their environment is essentially similar to that of their ancestors, which means that the toolkit they carry at birth is appropriate to their current circumstances.

All life needs to be fuelled by a source of energy and, for all practical purposes, that energy source is the sun. Lifeforms that capture solar energy by photosynthesis are bacteria; single-cell organisms such as algae and plants that contain chlorophyll. Aquatic lifeforms that capture solar energy need access to light so lie close to the surface. The generic name phytoplankton is used to define those organisms that capture solar energy in the seas, lakes, rivers and ponds. There are at least 5000 recognised

Animal Welfare: Understanding Sentient Minds and Why it Matters, First Edition. John Webster.

species of phytoplankton, which is interesting but unimportant in this context. What is important is the sheer magnitude of the food resource that it provides for animal life in the waters. It has been estimated that the phytoplankton of the oceans capture about 50% of the cardon dioxide and produce about 50% of the oxygen for the planet.

Box 6.1 presents a somewhat idiosyncratic summary of the food chain in the waters. At the bottom of the chain are the producers; organisms that may be loosely classed as 'vegetables' because they capture solar energy by photosynthesis. The phytoplankton are by far the dominant contributor to this category, but it also includes larger, plant-like seaweeds and seagrasses that grow abundantly in shallow water near the shores.

Essentially all other forms of sea life are animals: an astounding variety of species ranging from the polyps that construct coral reefs to the great whales. They may be classified, rather imprecisely, into herbivores, foraging carnivores and hunting carnivores. The great majority of herbivores are the zooplankton, small invertebrates such as molluscs, jellyfish and small crustacea that feed on phytoplankton. Large herbivores include fish (e.g. carp) and mammals (dugong and manatee). The feeding behaviour of the carnivores may be described, somewhat loosely, as foraging, hunting or a combination of the two. I use the term foraging to describe behaviour that involves the relatively mindless harvesting of small creatures from the waters (e.g. zooplankton) and seabed (e.g. molluscs). Foragers, by this definition, include creatures large and small: fish, crabs, some pinnipeds (e.g. walrus) and baleen whales. Hunting carnivores are predators; animals that select their prey and kill them. This category includes fish (e.g. tuna, sharks), birds (penguins) and aquatic mammals (e.g. seals, dolphins). When we exclude single-cell organisms, survival strategies for nearly all sea creatures depend on killing and eating other animals while avoiding being killed and eaten themselves. The question to be addressed here is how much of this behaviour is instinctive and how much may require properties of a sentient mind?

The nature and life of the open oceans has remained reasonably constant for millennia, although it is now under threat. There are, of course, large variations in the physical

Box 6.1 The Food Chain in the waters

Producers: 'Vegetables' that capture energy by photosynthesis
- Phytoplankton: single cell organisms, bacteria, algae
- Seaweeds: multicellular plant-like organisms: kelp

Consumers:
- Herbivores:
  - Zooplankton: molluscs, jellyfish, small crustacea (shrimps)
  - Herbivorous fish (carp), mammals (dugong, manatee)
- Foraging carnivores
  - Small fish that feed on zooplankton
  - Cephalopods (squid), large crustaceans (e.g. lobsters)
  - Mammals: walrus, Baleen whales
- Hunting carnivores:
  - Large fish: sharks, tuna
  - Aquatic birds: penguin, puffin
  - Aquatic mammals: dolphins, orcas, seals, walrus, sea otters

properties of the sea, like temperature and salinity, determined by latitude and prevailing currents, and these determine the distribution and abundance of life within the food chain. Nevertheless, the fundamentally stable character of the ocean environment, and its lack of surprises, greatly reduces the need for animals to develop new skills and better understanding in order to sustain fitness. Aeons ago, the shark evolved into a perfect killing machine and has been under no pressure to evolve further, or even give much thought to the business of life and death. However, if smaller fish, at risk of being killed and eaten by sharks and suchlike, are to achieve Darwinian fitness (the capacity to survive and reproduce) they need to develop defence strategies. As always, they are motivated by the emotional need for self-preservation. This requires at least some instinctive notion of the concepts of pain and fear reinforced, in many cases, by some acquired understanding of actions necessary to keep out of danger.

While the oceans themselves may be relatively uniform, the environment of the seabed is complex and presents a wide variety of environmental niches that present complex challenges that encourage the development of special skills, including properties of a sentient mind. These include sheltering as a form of defence, camouflage and (perhaps) deceit as aids to attack. Further demands for navigational skills are placed on marine animals whose breeding cycle requires them to migrate between sea and land, like the sea turtle or between fresh and salt water like salmon and eels.

There is clear evidence of properties of the sentient mind in a wide range of marine animals. We are fed many tales of the intelligence and sensitivities of the cetaceans (whales and dolphins), descendants of ancestral mammals that returned to the sea to exploit an abundance of food. It is easy to empathise with these animals because they seem to be so similar to us. It may be harder to empathise with fish and marine invertebrates, but there is abundant scientific evidence to show that most fish and many invertebrates, including probably crustacea (e.g. lobsters) and certainly the cephalopods (octopus and squid) display a range of emotions and acquired skills indicative of the deeper circles of sentience (69). Here, I examine attributes of mind ranging from pain and fear through to empathy and compassion in marine animals ranging from the octopus to the whale with no attempt to pre-classify them as inherently more or less sentient or skilled.

Pain and Fear

In Chapter 2, I drew attention to the distinction between nociception, a sensation such as an electric shock that produces an immediate withdrawal response, and true pain as a source of suffering resulting from a complex mixture of unpleasant sensation, emotion and fear of its recurrence. The extent to which pain can be more than just nociception comes from evidence of avoidance behaviour, guarding wounded sites from future trauma, changes in mood such as increased fear or depression, improved mobility and mood after administration of analgesics and, where this can be tested, self-administration of analgesics as a conscious action intended to reduce suffering. We probably assume that cetaceans (whales, dolphins etc.) suffer pain and fear in much the same way as we do, not least because their brain development is similar to ours so there have been few serious attempts to find out for sure. In a moral sense, it is more important to discover how far the capacity to experience pain and fear as a form of suffering may extend to the fish and other marine creatures, including invertebrates, that we hunt, farm and kill for food. For many years, it was firmly believed by most fishermen and too many scientists that fish do not feel pain. Fishermen made their case on the basis that a hooked fish makes strenuous efforts to escape and claimed that it would not do this if it hurt. This argument ignored evidence of trapped foxes and mountaineers respectively gnawing or sawing off trapped limbs in order to survive. For many years, scientists argued that fish and marine invertebrates like crustacea and cephalopods could not feel pain because they do not possess the area of brain known to be associated with pain reception in mammals. Fishermen and scientists alike came to this conclusion without once asking the fish (or, indeed, the squid). There is now convincing evidence that both parties were wrong. Fish do suffer from both pain and fear.

Bony fish meet most of the criteria necessary to classify pain as a source of suffering (71,72). They avoid areas they associate with electric shocks, although (reluctantly?) will visit these areas if, after several days of starvation, they are the only source of food. The movement of injured fish (sensation) and their motivation to feed (mood) is improved by the administration of analgesics, and there is some evidence that they will select a barren environment in preference to an enriched environment if it contains an analgesic in solution.

Fishing with hook and line induces both a painful stimulus and the fear of being trapped. Experimental studies have compared the behavioural and biochemical stress responses of fish to the administration of small electrical shocks to the mouth (pain only) with their responses to being hooked and reeled in (pain and alarm). Their stress responses to the latter were far greater, indicating the far more severe combined stresses of pain and fear. At this point, I reprise my opening paragraph (and why not? It's a good story). Some years ago, I was involved in a live debate on fishing and fish welfare before a late-night studio audience. After I had outlined the results of these experiments, a member of the audience said 'This is all rubbish. These scientists don't know what they are talking about. I have been fishing all my life and I know for certain that fish don't feel anything'. He then added 'What sort of fish were they anyway?' and when I said 'carp' he said 'Ah well, they're clever buggers'. While this remark was a perfect illustration of our sloppy thinking in regard to animal sentience, and its implications for our conduct, he may have unintentionally brushed the edges of a possible truth. We cannot assume that all fish feel pain in the same way. It is possible that sharks and other cartilaginous fish may not suffer pain. They bite each other during mating rituals and cause quite major injury but this does not appear to affect their behaviour at the time or thereafter.

Prey species of fish that inhabit the oceans occupy a similar niche in the food chain to herbivores (impala, wildebeest) that inhabit the Savannah. Both have adapted to the challenges of life through patterns of behaviour that, ensure, so far as possible, that they are unobserved by predators (through sight, sound or smell) or at a safe flight distance. In these circumstances, it is reasonable to assume that they feel reasonably secure. Moreover, they are likely to be killed the first time they are caught, so unlikely to live with bad memories of previous capture. There is, of course, an exception to this rule. Fish that become unwilling participants in the sport of angling may be caught several times, stored in keep nets, where they are liable to suffer more injury that during the catching process, then returned to the water. One might expect them to learn from this process. The evidence for this is scanty, which is hardly surprising, since it must be difficult for a fish to distinguish a baited hook from other food sources. Nevertheless, some reports suggest that carp that may live for over 20 years become, over time, more suspicious and harder to catch. Learning from long experience, they may indeed grow into 'clever buggers'.

We have clear evidence of nociception (at least) in cephalopod molluscs (squid, octopus, cuttlefish) and crustaceans (crabs and lobsters). There are suggestions of more complex reactions to painful stimuli in crustaceans, including avoidance behaviour and response to analgesics, but here the evidence is scanty. However, cephalopods such as the octopus and squid display many of the signs we would associate with pain as a source of suffering, measured in the context of sensation and mood (14, 69). They learn to avoid potential sources of pain, for example, sites associated with electric shocks. When one limb has been injured, they avoid using it and sometimes carry it around in another limb to keep it away from painful knocks. Pain is a very subjective experience, and I cannot even be sure how my pain would feel to you, my reader, another human. However, if we accept the precautionary principle, we must assume that bony fish and at least some of the invertebrates that we catch for food or sport meet at least the first three skandhas of sentience, which means that pain can be a source of suffering. The Animal (Scientific Procedures) Act., UK (24) includes fish and cephalopods in the category of animals requiring protection in the matter of procedures calculated to cause pain, suffering, distress or lasting harm. The number of species guaranteed this protection (e.g. at least some of the crustacea) may have to be increased in the light of new knowledge. Cephalopods have a complex, hierarchal nervous system, which is not present in other 'simpler' molluscs such as oysters, mussels and scallops. However, we cannot exclude the possibility that these too, may not only sense but perceive pain.

Survival Skills: Hunting, Hiding and Problem Solving

Most predation of smaller fish by larger fish (e.g. tuna and sharks) in the open ocean appears to involve little, if any, thought from either party. Sharks are selective, they don't just snap at anything that moves, and, when they are hunting the seabed where there are places for small fish to hide, they know where to look, but I think that's about it. Shoals of prey fish swarm when under attack but this behaviour looks to be instinctive, each individual moving according to a programmed set of rules. The complex environment of the seabed does, however, offer a wide range of opportunities and tools to both predator and prey. However, most hunting and hiding strategies are likely to be instinctive. To quote but two, from many examples. Flat fish, like plaice, can hide in the sand; their dorsal markings rendering them almost invisible. Angler fish can lie still, holding out their lures to attract unsuspecting fish almost into their mouths.

My concern here is to explore evidence indicative of higher intelligence, or cognitive ability in the hunting and hiding strategies of aquatic animals. This may involve tool use, problem solving and the acquisition of knowledge and understanding by observation and interpretation of received images. Aquatic mammals, unsurprisingly, employ a range of advanced skills. Orcas hunt in packs using strategies that depend on communication and coordination and these improve with practice. There is, to my knowledge, only one convincing example of tool use in wild cetaceans. Bottlenose dolphins in Shark Bay, Australia, carry sponges around with them to protect their noses when foraging in the sand. This behaviour is a local phenomenon, rather than a property of the species, so appears to be culturally acquired (8). Some fish build shelters from discarded shells and other convenient materials lying around on the sea floor. This may be a hard-wired expression of a nest-building instinct, akin to that of the freshwater male stickleback who builds a nest before attracting a female to spawn inside it.

The fish with the most skilled feeding strategy is the archerfish that knocks insects off branches overhanging the water by spitting at them (see p. 49). Much of it is hard wired. The elements of higher intelligence include calculating the quantity of spit according to the height of the branch, predicting where the insect will fall, and developing through constant practice, strategies that reflect the individual, not just the species. The squid that gathers fallen coconut shells and carries them around as disguise, both for defence and attack appears to be practising deceit.

These observations carry conflicting messages. The first simply confirms the evidence that tool use is not strictly confined to 'higher' mammals such as the primates. The second message is that the ability to use simple tools is not necessarily an indication of higher intelligence. It could be a simple evolutionary response to the primitive need to eat in an environment where much of the food, e.g. shellfish, is only accessible after the can has been opened. For a pattern of behaviour to be classed as an indicator of higher intelligence, it must include an element of learning. The octopus, which is a solitary creature, probably improves its hunting skills through trial and error, as does the archerfish. I suspect (but don't know) that wrasse learn how to open scallop shells from observation of others. Generally speaking, social learning tends to be more highly developed in species like the dolphin in which social behaviour involves elements of deep sentience such as affiliative behaviour and possibly theory of mind.

It is difficult to demonstrate evidence of problem-solving skills in wild animals living in their natural environment. Two species that have been subject to extensive study under laboratory conditions are the bottlenose dolphin and the octopus. The octopus can learn to navigate a simple maze and pick up other skills by observation. Octopus in laboratories also learn, without instruction, to climb out of their tanks, slither over to another, steal a crab and return home before they are discovered. Many of the studies with dolphins have sought to discover how closely related their minds are to human minds as measured by properties such as understanding of numbers and signals, self-awareness and theory of mind. Most captive dolphins participate willingly and do some sensible things in these experiments, but firm conclusions are hard to come by. Moreover, as I have written before, I am not convinced that the ability of a sentient animal to interpret and respond to questions appertaining to human consciousness is a particularly useful measure of the extent to which intelligence contributes to its fitness and wellbeing in its own natural environment.

Migration

Many marine animals migrate long distances for much the same reasons as migrating birds; namely between areas where food is most plentiful and areas where it is more likely, or even essential to achieve breeding success. Most whales, especially the baleen filter-feeders, migrate between feeding grounds rich in plankton in the cold high latitudes and selected, warmer, coastal waters closer to the equator, where they give birth and rear their young in locations, typically shallow waters where they are relatively safe from predators, especially orcas. They are able to judge the depth of the water by echolocation, measuring the time between their own call and the returning echo from the sea bottom. Seals and walrus have to return to land (or ice flows) to give birth and feed their young for the first few days or weeks of life. Sea turtles also return to known safe beaches to lay their eggs. These things require a high degree of precision in navigation. Salmon and eels do not care for their young, but they must spawn at the only sites where the young will survive. Atlantic salmon migrate between rich feeding grounds, e.g. off the coast of Greenland and Labrador and their spawning grounds in the clear, fresh, shallow waters of the mountain streams of Northern Europe. Eels migrate in the opposite direction. They spawn in the near stagnant, vegetation-rich waters of the Sargasso Sea and spend adult life, which can last for many years, in and around the brackish and fresh waters of rivers. In both cases, the movement of the very young is almost entirely passive. Salmon eggs hatch into alevins that remain upstream, obtaining nutrition from their yolk sac, then mature through stages somewhat arbitrarily defined as fry and parr as they move down to the salt waters of the estuaries. Their net movement downstream is driven by the current of the river, but they do spend time swimming upstream, which helps to delay their exit to the sea until they are physiologically and behaviourally able to cope with life in salt water. Thereafter, they migrate with no special urgency to the feeding grounds. Neonatal eels (leptocephalus) have little or no control over their subsequent direction and are swept north-west by the Gulf stream and North Atlantic current to end up as tiny glass eels in the brackish and fresh waters of the rivers of northern Europe, where they mature over many years through various stages into mature silver eels. The motivation of adult salmon and silver eels to return to their breeding grounds is driven by powerful, hard-wired needs that override nearly all the subtler aspects of sentience, so fall outside the scope of this investigation. However, the success of these planned journeys to specific breeding sites requires some very special skills in navigation.

Migrating fish and other sea creatures (e.g. cetaceans, sea turtles) cannot navigate from conspicuous landmarks. They may get some information from the sun, although this would seem to be limited. They have the faculty of magnetoreception, possibly in the form of a three-dimensional magnet that can, at least, guide their direction of travel relative to magnetic north and may also give them an indication of latitude by detection and interpretation of the earth's geomagnetic fields. We think that as salmon near their breeding grounds, they sense the smell (or taste) of home and, as they approach, the increasing intensity of this specific and well-remembered smell guides them to the mouth of their home river (or occasionally, one nearby). The return of adult eels to their breeding ground in the Sargasso Sea remains one of the great mysteries of this mysterious species. Why, after an indefinite period ranging anywhere between 5–25 years in the waters and mud of the Somerset levels, should they suddenly decide to leave? When they do, how do they get there? Researchers have, so far, failed to track their route – tags slip off. They probably do not take the direct route but a more circuitous, easier path that follows the prevailing currents of the sea. Currents flow round the still waters of the Sargasso Sea in a clockwise direction. The north-east current at the top carries the miniature eels willy-nilly into the Gulf Stream and thence to the rivers of northern Europe. Currents flowing to the south and west at the bottom help the adults to return. However, in order to obtain the benefits of the prevailing currents, the eels appear to plan their course by swimming first south then west. Their navigational strategy has much in common with that of small boat sailors seeking an easy passage from northern Europe to North America, which is to 'sail south until the butter melts (somewhere off Portugal) then turn right'.

The navigation of sea mammals over relatively short distances is remarkably accurate. Seals captured off their home territory of the Bass Rock, in the Firth of Forth (Scotland), tagged with a radio locator and taken 200 km away into the North Sea, head unerringly straight for home when released. Once again, the special senses involved are likely to be magnetoreception and smell.

I cannot leave the subject of navigational skills in sea creatures without drawing attention to the extraordinary phenomenon of the mass stranding, or beaching, of whales. Records indicate that about 2000 cetaceans (whales and dolphins) are beached every year and unable to return to the sea without assistance. Stranded individuals may have been sick, injured or even died at sea, and we cannot necessarily attribute these to failures in navigation. There are, however, many records of mass strandings of cetaceans dating back many thousands of years. In 1902, about 1000 whales were stranded on the Chatham Islands, east of New Zealand. In recent years, there has been concern that ships using powerful sonar (echolocation) equipment for submarine detection can lead to strandings. The nature and volume of sound produced by this equipment can interfere with the whales' own echolocation and may be so loud as to damage their hearing. However, reports of mass strandings of whales long precede the development of sonar. Moreover, they have been recorded in only half the species. All of these are toothed whales, none are baleen filter-feeders. The link between sonar signals and mass strandings has only been observed in one species, the beaked whale. Strandings usually occur on shelving beaches, and it has been suggested that the animals fail to identify the gradual decrease in depth by echolocation. Cetaceans adapted to life in the shallows rarely get into trouble. However, none of these factors can explain the phenomenon of the mass stranding. These occur, rather obviously, in the species that live in large social groups, but this does not explain this mass exercise in self-harm. One suggestion is that one or more individuals gets into difficulties, sends out alarm calls, others rush to their aid, and they all end up on the beach. While it is not possible to come to any firm conclusions as to the reasons for this rare phenomenon, it does suggest that the process of navigation in cetaceans is hard-wired, operates on the basis of a few simple cues, and involves little, if anything, that could be called deep sentience in the form of learning, education or culture.

Communication and Social Behaviour

It is necessary to make the distinction between animal species that simply go around in groups and those who live a truly social life involving complex forms of communication, stable families, distinction between strangers and neighbours, and possibly displays of empathy and compassion. Many species of fish travel in shoals in pursuit of things to eat and to reduce the risks of getting eaten. Their behaviour may reflect some properties of their social life but, in the absence of evidence to the contrary, we may assume that each fish is operating according to a set of simple cues that have evolved to promote the success of the species.

Pinnipeds (seals and walrus), while at sea, feed mostly as individuals. However, they need to leave the water in order to breed on land or the ice so, at this time, they are compelled to congregate. The young are born covered in fur, which is an excellent insulator against the cold on dry land but useless in the water, where the insulation needs to be under the skin in the form of fat. The energy cost of producing a fur-lined baby is much less than the cost of producing a fat baby. Seal pups are fed a very rich milk for a very short time, sometimes no longer than three weeks, during which time, they readjust their insulation from outside (fur) to inside (fat), at which time they are ready to go to sea. Social life during the brief breeding season involves little, if anything, in the way of cooperation or companionship. Males, most conspicuously the walrus and elephant seal, which contribute nothing to the rearing of their offspring, engage in brutal and bloody battles for control of their harems.

Most cetaceans, dolphins, porpoises and whales, are highly social creatures that rely primarily on sound as a means of communication. They are sensitive to smells from far-away sources but probably unable to determine their exact location. They use high-frequency sound signals, clicks and whistles, to transmit specific information at close range. Low-frequency sounds, songs and moans, travel a long way, quickly. (The speed of sound in water is five times greater than in air). The two most studied species of social cetacean are the orca (killer whale) and the bottlenose dolphin, not least because both species have been exploited as attractions in theme parks. Orcas live in extended family groups. Young males, in particular, may stay close to their mothers throughout their lifetime, which may be as long as 50 years. They use whistles to communicate with each other within the band. Individual dolphins and orcas have their own signature whistle, which is recognised by others. Moreover, an orca or dolphin that wishes to communicate with a chosen companion can call it by its name (31). Clicks are used by dolphins, orcas and other toothed whales for echolocation. Like bats, they can identify the size, distance, direction and speed of movement of prey (or strangers) from the delay, magnitude and pitch of the returning echo. Low-frequency sounds, classically the 'song of the humpback whale', that travel fast and far, are used to locate and attract sexual partners. Whales in different locations sing different songs and these songs change over time, which implies that musical performance in whales, as in birds and humans, is largely a product of learning and mimicry but does offer scope for the imagination.

Dolphins play. The way they play would seem to have little to do with the acquisition of essential life skills and everything to do with the pursuit of pleasure. Dolphins surf the bow waves of boats and ships. In my personal experience of the sea, they appear more likely to join me when I am sailing silently than when motoring. Groups of close companions dance together and share their toys (56). Most curiously, groups of dolphins appear to 'play' with pufferfish. When attacked, this fish releases a toxin that, diluted in sea water, appears to induce a narcotic or euphoric state in the dolphins, who learn to come back for more. In short, dolphins can get high on puffer fish. Expressed in more sober scientific terms, they acquire a taste for a pleasure that is not directly linked to food.

Dolphins and orcas in aquatic theme parks ('Sea Worlds') can be taught to do elaborate tricks for food rewards. While this behaviour is evidence of advanced mental formulation it is not spontaneous so cannot be classified as play. Moreover, there is good evidence that captive cetaceans first become bored, then stressed by the constant demand to entertain. In this example, demonstration of the ability of a confined, non-human species to perform tricks thought up by humans is not only irrelevant to the understanding of their behaviour in the wild; it is also cruel.

Dolphins in the wild live a loosely structured and reasonably relaxed social life within which they recognise friends, neighbours and strangers. Male dolphins in particular form alliances with two or three close friends, which they reinforce with displays of affection like nose rubbing, and mutual play. The size of the larger social group of neighbours varies between pods and over time. Good communication and cooperation within a substantial band of neighbours brings benefits both in terms of hunting strategies and defence against sharks. There is a hierarchical structure within these larger pods and competition for dominance within the pecking order can involve fights and injuries.

The behaviour of dolphins and other cetaceans justifies their inclusion among the group of animals showing the deepest degrees of sentience as measured both in terms of emotion and cognition. Most of the arguments in support of special status for cetaceans also make reference to their brain size but, as I wrote in Chapter 2, I avoid reference to brain structure and size because it completely fails to account for, for example, the problem-solving skills of corvid birds, the emotional response to pain and fear in fish, and the complex hunting and hiding strategies of the cephalopods. I say again: detailed study of the form and function of the brains of specified animals can constrain our capacity to understand the minds of animals very different from ourselves. I suggest it is better to look for evidence for three strong indicators of deep sentience: grief, empathy and compassion. Several species of dolphin and one species of whale (the humpback) display what scientists call 'post-mortem attentive behaviour' (3). This involves staying in close physical contact with a dead neighbour. Female dolphins have been seen carrying their dead calf around for weeks while it progressively disintegrates. This looks like grief. It could just be a failure to recognise that they are dead. Measurements of stress hormones would help our understanding but these, unsurprisingly, are not available. One of the best positive expressions of empathy and compassion is the decision to go to the aid of another animal (not one of the family) who may be in trouble. Dolphins form defensive alliances when under threat of attack from sharks, but this could be interpreted as cooperation motivated by self-interest. There is however good evidence, including video evidence, of a pod of dolphins forming a defensive wall around human swimmers when sharks are in the vicinity. There are many stories, dating from early history, of dolphins coming to the aid of swimmers in distress and, once again, some new video evidence to support these tales. We cannot be sure, but I think we may take it as highly likely that the behaviour of some social species of dolphins is consistent with the principles of empathy and compassion, so may be given as further evidence that the deepest circle of sentience, namely consciousness, is not a property unique to the human species.

Animals of the Air

This chapter ponders with amazement on the animals of the air. The ability to escape the surly bonds of earth gives birds a degree of freedom from human interference, not open to land mammals. The only species that have succumbed to domestication are (almost) flightless birds such as chickens and turkeys. The ability to fly carries great advantages in respect to essentials of life like foraging for food and escape from predation. However, birds need to keep their weight down. With a few exceptions therefore, birds are better equipped for flight than fight.

While this chapter may be short on answers, it asks the same questions as those critical to our understanding of the minds of any sentient species. What are the things that matter most? What are the features of the physical and social environment that determine these priorities? What are the skills and properties of the sentient mind that enable them to cope with the challenges of the environment? We must begin, as always, with a restatement of basic principles. All lifeforms, plant and animal are designed and operate to meet their basic needs and promote their genetic fitness and much of this is instinctive. Basic needs include access to sufficient food and water to sustain life, protection from the elements, safety and security from predation and breeding success, Animals with one or more of the powers of deep sentience; perception, mental formulation and possibly consciousness, can build on their mental birthright and develop their minds through a combination of experience, learning and practice.

Animal Welfare: Understanding Sentient Minds and Why it Matters, First Edition. John Webster.

Birds, direct descendants of reptiles, have adapted to every ecological niche on earth, water and the shorelines where they meet. The variety of habitats inhabited by birds and the range of strategies they have developed to deal with the challenges presented by these habitats is vast. Our fascination with these abilities has generated an immense amount of scientific investigation into bird behaviour, motivation and special skills, to which this chapter cannot begin to do justice. Here, I restrict myself to examination, by way of a few examples, of how much the success of birds in meeting their basic needs is built into their intrinsic genetic birthright and how much may be linked to deeper cognitive and emotional aspects of sentience. These include complex mental formulations such as navigation and problem solving, communication and social skills, and higher emotions such as empathy and compassion.

When each individual bird leaves the protection of its parents, it is programmed to find a supply of food appropriate to its design for eating and digestion, find a mate with whom to reproduce, select a breeding zone and individual nest site that will combine security with access to a source of sufficient food for all, then work hard to provide food and instruction to give the offspring the physical ability and mental skills necessary for independent survival. As parents, birds are powerfully motivated to do their best for their brood. There will be many failures but, so long as the population is maintained, the exercise will succeed.

Feeding Strategies

Birds are commonly classified according to their feeding habits as:

- Carnivores: raptors, eagles, hawks and other birds of prey
  - carrion eaters: vultures, carrion crows
  - fish eaters: pelagic birds, albatross, Arctic terns.
- Frugivores: fruit eaters, e.g. orioles, waxwings
- Granivores: seed eaters, e.g. sparrows, tits, finches, chickens
- Insectivores: swallows, swifts, flycatchers, warblers
- Mollluscivores: waders and other shoreline birds
- Omnivores: gulls, blackbirds, chickens

This classification is far from definitive. For any species of bird to succeed, it must either have continuous access to a wide range of food sources that it can grasp, swallow and digest and/or preferential access to a food supply that is inaccessible, or less accessible to other competing species. Darwin's finches on the Galapagos evolved different shaped beaks that enabled them to harvest different fruits, nuts and insects, thereby exploiting the full range of food on offer with minimal competition. Birds of prey are designed and motivated to hunt and kill. Waders and other birds of the shoreline like the avocets, dunlin, oystercatcher and spoonbill, have evolved distinctive beaks adapted to specialist feeding behaviour. Avocets and spoonbills sweep the shallows for small aquatic creatures, dunlin and oystercatchers peck around the tideline for molluscs and other small shellfish. Swallows and swifts have adapted their bodies to a life spent mostly on the wing in order to harvest the abundant crop of flying insects. While these specialisations give preferential access to a particular feed source, they limit the range of foods on offer. When that supply is seasonal, birds with specialised feeding habits have to move with it. Swans and geese that feed mainly on grasses and other vegetation, summer in the subarctic but migrate south for the winter. Swallows winter in the tropics and migrate north for the summer to breed at a time when the insect population is at its highest and the days are long.

It is self-evident why geese migrate south (or to warmer climes) for the winter. It is not immediately self-evident why swallows should migrate north for the summer. The tropical forests should provide a plentiful year-round source of insects. The Darwinian explanation for migration behaviour is that a subset of birds within certain species, e.g. insectivorous species like the swallow, migrated north while the remainder of the population retained the easier option of a settled life in the tropics. The migrants succeeded because, on average, they raised a larger number of offspring. Likely reasons for this include longer daylight hours to give them more time to gather food, less predation of their nests and the probability of raising a greater number of clutches during the summer months. This made them the fitter, so they survived, so now all swallows migrate. Migration is a striking example of the fact that the successful option, measured in terms of the fitness of the population, is not necessarily the easiest option as might be chosen by the sentient individual, motivated by the desire for a life of comfort and security involving as little work as possible. The motivation to migrate appears to be hard-wired. The strategies involved in planning the journey require some very impressive properties of mind.

Those birds that do not migrate are those that have year-round access to food that they are adapted to eat. Birds of prey (raptors) have access at all times to smaller birds and mammals like mice and shrews. However, their numbers are constrained by the numbers of their prey. If the numbers of prey decline, or the number of raptors increases after a particularly successful breeding season, individuals will go short of food. The first consequence of a reduction in food supply for most animals is usually a fall in the reproductive rate. Adults may survive but produce fewer chicks. It follows that we shall never be overrun by raptors.

The most successful of the wild birds are the opportunist omnivores like gulls and crows. It should come as no surprise that they are also extremely smart, since they have devised a wide range of strategies for getting food without recourse to the rigours of hunting or migration. In Chapter 4, I described the range of skills demonstrated in laboratory studies on developing complex strategies for getting food from flasks, seen but out of reach. Crows in the wild devise a wide range of strategies to get at inaccessible food, like the meat inside molluscs. These include relatively primitive approaches like hitting a mussel with a stone, dropping it from a height on to a hard surface and, my favourite, taking them to a car park, then carefully placing them at a spot where they know they will get run over by a car tyre.

The herring gull is a bird that has achieved spectacular success as an opportunist omnivore without the need to give the matter much thought. Their name, herring gull is another example of our penchant for describing species by how we see them, not as they are. Sure, gulls follow fishing boats to pick up what is thrown overboard because it is easy pickings. When there were masses of fishing boats, that was the best place to be. As the fishing fleet grew fewer in number, the gulls migrated to the rubbish tips.

Now we can recycle our edible waste, gulls are abandoning the rubbish heaps and migrating to the locality of fish and chip shops and other takeaways. Two elements of emotional, rather than cognitive intelligence have contributed to their success. The first is a brazen approach based on the confident assumption that a human eating fish and chips is most unlikely to do them harm. The second is more subtle: the ability to steal up on a food source, e.g. a baguette in a shopping bag, without being noticed. I have often watched this behaviour and am inclined to believe, without proof, that it involves not just stealth but a measure of deceit. Herons display a particularly advanced strategy for improving access to food, namely the use of ground bait to attract fish. They have been observed to entice their prey with a wide range of ground bait, seeds, feathers and scraps of abandoned foods. Individuals pick up this skill from observation of others, not necessarily their parents. This strategy requires an advanced property of mind, namely the understanding required to work for an indirect or delayed reward.

All these examples of deep sentience and advanced skills in birds relate to the motivation to get a reward that really matters in the context of survival, i.e. food. The question arises: 'are birds motivated to employ these advanced skills in trivial pursuits such as play'? Parrots peck away at toys in their cages, but what we call play may be an expression of frustration rather than pleasure. Keas steal from picnics and vandalise cars. This would appear to give them satisfaction but can hardly be described as imaginative. The most convincing display of play behaviour that I have witnessed (if only on YouTube) was that of a crow (surprise?) in Russia. This bird had picked up the lid of a jam jar and was using it to toboggan repeatedly down the sloping side of a snowy roof. It was clearly able to distinguish between the snow-covered, good sledging zone and the bare surfaces that weren't. When it had dislodged much of the snow, it flew off, still grasping the jam jar lid, maybe to continue the fun elsewhere.

Migration

Migration is essential to the fitness and survival of many species to ensure year-round access to their food supply and a secure and well-stocked breeding site. The migration programme is hard wired and begins several weeks before take-off. Geese that feed on land but migrate long distances over water increase food intake and lay down large quantities of body fat as fuel for the journey. Changing daylight length triggers hormonal changes that reset the appetite control mechanism several weeks before the date of take-off. Most migration patterns in birds of the land and shorelines involve north-south movements between the summer breeding grounds at high latitudes and winter-feeding grounds in the tropics and subtropics. Some species, e.g. hummingbirds in North America, will choose not to migrate if there is still plenty of food at home. Some birds migrate very long distances. The most spectacular of these is the Arctic tern, which migrates from its summer breeding grounds in the Arctic to seek a second, non-breeding summer in the Antarctic. The shortest distance between the two sites (as the crow flies, but not the tern) is about 20 000 km. However, terns identified by tracking devices have been shown to make journeys up to 80 000 km, flying first down the west coast of Africa, then south-east across the south Atlantic to reach the Antarctic continent south of Australia. It has been claimed that they choose this route to take advantage of the prevailing winds. I suggest that when they hit the Roaring Forties (strong westerly winds of the southern hemisphere), they have little option but to go with the flow. The distances travelled by the Arctic tern may seem extraordinary to us: over a lifespan of perhaps 20 years, these journeys can add up to several trips to the moon and back. However, these distances appear to present no problem for the tern. Food is always available at sea, although more plentiful in the colder waters. Their flying/gliding pattern is very energy efficient, and it is easier to glide downwind than fight the elements in an attempt to maintain a track of due south.

The albatross is often described as a migratory bird. Couples mate for life and return to the same nest to rear a single chick every 2–3 years. In the years when they are not rearing a chick, they circumnavigate the Southern Ocean, flying and gliding downwind. If migration is defined as long-distance travel in search of food, the behaviour of the albatross is the exact opposite. For a bird that feeds on food from the sea, one bit of the Southern Ocean is just as good as any other. However, in the years when they breed, they are committed to return to the nest for over 200 days, first to incubate, then to feed their single chick. This is much harder work. Not only do they have to forage for more food, but they have to fight the prevailing westerlies in order to stay close to the nest.

While the motivation for migration and other forms of long-distance travel, like the homing instinct of the pigeon, may be hard wired, success on the journey requires special skills, some innate, many acquired. Oddly enough, the behaviour of the Arctic tern may be among the most primitive. If they leave Greenland and set a fixed course south, they will eventually end up in the Antarctic. However, conditions of wind and weather will ensure that their track (the direction they actually travel) will take a highly indirect route.

Most migrating birds don't just set a course for their destination and fly in what appears to them to be a straight line. Migrating land birds (and butterflies) establish flightpaths that select the shortest possible sea crossings, e.g. across the Straits of Gibraltar. Birds that are blown off course find their way back. Migrating shorebirds such as waders that have to travel long distances over land know the location of a number of suitable feeding stops *en route*. For this, they require advanced skills in navigation, usually reinforced by the memory of those in the flock with prior experience of the route.

Migrating birds have the faculty of magnetoreception that, as described earlier, enables them to set a course in relation to magnetic north and may give them some indication of latitude (32). They can steer by the sun (and possibly the stars) and those that fly repeat journeys over land will recognise familiar landmarks. In Chapter 4, I compared the navigational equipment of migrating birds with that available to early ocean voyagers. Cabot and Columbus would have had a reasonable idea of where they were (latitude better than longitude) and where they were heading but no real idea where they would end up. Later voyagers had the benefit of charts. Most migrant birds are aiming for a very precise location. For this, they need to know the coordinates (latitude and longitude) of both their destination and where they are at any point of the journey. Somehow, they manage. We know that migrating birds, blown seriously off course, or deliberately moved hundreds of kilometres from their flight path, can reset their bearings and successfully reach their planned destination.

We have good evidence that homing pigeons, through personal experience and observation of their better-informed colleagues, can build up detailed maps of wide areas around their home base from observation and recall of conspicuous landmarks such as radio masts and church steeples (7). However, it requires too great a stretch of the imagination to assume that each swallow comes equipped with a detailed chart of the northern hemisphere. Moreover, even if this chart clearly marks the points of departure and arrival, it cannot tell the bird where it is at any stage of the journey and thereby what course adjustments may be necessary in order to reach its final destination. Birds that migrate mainly over land tend to migrate in flocks: those on their first flight flying with the experienced adults, who probably have the memory of conspicuous sights *en route* such as the presence and shapes of mountains and coastlines. This does not explain how birds can migrate successfully long distances over the featureless oceans. We don't know how migrating birds are able to plot their journey coordinates (latitude and longitude) with sufficient accuracy to meet their needs. This requires special senses that we humans do not possess and can scarcely imagine. As flocks of migrating birds approach their destination, they are able to home in very precisely on sights and smells familiar to the experienced travellers and these may arise from a long way off. However, this does not begin to explain how an albatross, having traversed vast distances of the Southern Ocean, can return to the same nest on the same small island it left two years previously.

To summarise: the initial motivation to migrate is instinctive in most birds, although some, like the hummingbird, are known to change their minds. It may be significant that the hummingbird, unlike, for example, the Canada goose, cannot lay down a large store of fuel before setting out on its journey, so has evolved a more thoughtful approach to decision making. Long-distance navigation, especially over featureless regions of the ocean, must depend on hard-wired skills based on magnetoreception, the direction and elevation of the sun, and time of day. These skills may be instinctive, but they are of a very high order. The ability of the pigeon or swallow to find its way to the exact spot that it calls home depends on the skill that we call orienteering (navigation by map reading). This advanced mental formulation is acquired through learning based on personal experience and observation of the elders, accompanied by an astonishing ability to memorise large, detailed maps.

Those who dismiss the notion that birds are able to create and retain mental images that far transcend most human understanding may care to reflect on the fact that this property is shared by concert pianists, who can, by dint of intense practice, memorise hundreds of pieces, many consisting of thousands of notes. Others with this extraordinary ability to memorise every fragment of the big picture are individuals with the rare form of Asperger's syndrome that enables them, after a single glance, to draw in perfect detail a complex image like the west face of a Gothic cathedral. We have little idea how these minds work, but if it is possible for a human mind to store every detail of the façade of a Gothic cathedral, it is equally possible for a pigeon to store every detail of the map of south-west England. One can only be amazed.

Sentience and Breeding Behaviour

The genetic fitness of any animal species depends, above all, on breeding success. Each species has evolved a phenotype and developed a pattern of behaviour to meet the special demands of its habitat. In all species, breeding behaviour, both sexual and parental, is essentially hard wired: i.e. built into the birthright. The emotional basis of breeding behaviour was discussed in some detail in Chapter 4, with particular reference to breeding in birds. Here, I reprise the main points.

In altricial birds that feed their young in the nest, breeding success depends on hard work from both parents from the time they start to build a nest until some time after fledging. Male and female stay together and work together throughout the breeding season. The importance of this bond increases in proportion to the number of eggs laid and the amount of time and energy necessary to rear each chick. The most extreme demonstration of this is seen in albatross that rear one chick every three years. The genetic fitness of the albatross owes little to the urge to mate and almost everything to parental care. Precocial birds are mature enough to feed themselves from the moment of hatching but still depend on the hen for warmth, security and education. The number of precocial species is small and restricted to those adapted to life spent mostly on the ground.

While breeding behaviour may be hard-wired, the motivation of individual birds, and couples, to carry out any action is driven by an emotion. This applies both to courtship and mating and to parental care. All behaviours of parents in defence of their nest are highly charged with emotion, Some, like the broken-wing display or behaviour designed to attract a predator to a sham nest, demonstrate the advanced mental capacity to practice deceit (70).

Social Behaviour, Culture and Education

Sentient animals need to learn about life if they are to survive. The more vulnerable the individual, the more it needs to rely on observation of experienced parents and neighbours, rather than adopt a 'bull at a gate' process of trial and error. I repeat, you cannot afford to be brave if you are a chicken. The need to observe, interpret and interact with the behaviour of others assumes special importance in birds that live in social groups. Adaptation of a social bird like the chicken to a settled life in society calls for a much more complex understanding of interactions between neighbours than adaptation to life as a solitary, fiercely territorial bird like a kingfisher or robin. There is good evidence that an enriched social life enhances brain development in the individual, so it is fair to assume that it also favours the evolution of smarter species. Study of the social behaviour of birds is a massive subject, far beyond the scope of this little book. Here, I select a few examples to illustrate the extent to which individual birds can develop their minds and populations of birds can develop a culture from personal experience, observation of others and formal education. Much of this information comes from the observation of that much studied bird, the domestic fowl.

The social structure of a population of chickens is hierarchal and they can be aggressive in the way they maintain the pecking order. Different chickens express their personalities in different ways, and these personalities develop as they work out how they should interact with other individuals within the group (20). Aggressive feather pecking, leading in some cases to deaths, can be a problem in commercial flocks of laying hens and this has become a bigger worry as, in the pursuit of better animal welfare, more and more hens have been taken out of cages and reared on free range. Many studies of the factors that might contribute to feather pecking in hens have been carried out on experimental flocks with an average population size of about 60 birds and, in many of these studies, it was shown to be a serious problem. However, in a large study of hen welfare in well-managed commercial free-range flocks, often in excess of 10 000 birds, Becky Whay and I recorded no significant problems of feather pecking or aggression (80). We assumed that this was, in part, because they were widely dispersed in an enriched environment, However, the fact that the problem appeared to be less in very large flocks suggested to us that when populations were this large, birds found it almost impossible to recognise another bird as an individual, so were unable to establish a pecking order. Uncertain as to how to behave in the presence of strangers, both parties elected for caution.

Social learning, i.e., the regional accents of songbirds, the use of ground bait by herons, as described in Chapter 5, creates a culture wherein the behaviour of individuals confirms to the mores of the society. Moreover, it is clear that social learning is not just acquired by passive observation but enriched by the formal business of education. Some of this will be instinctive, e.g. the Southern Pied Babbler parents who prepare their chicks for fledging by training them to associate the arrival of food with a distinct call (60). However, the ability of bantam hens to pass on information acquired personally from experience gathered in their own lifetime, such as the belief that blue or yellow corn is poisonous (55), provides good evidence for the power of mental formulation (skandha 4). Some features of the social behaviour of rooks such as compassion for neighbours in distress, offer evidence of a degree of emotional sentience that may indicate entry into the deepest of the skandhas., namely human consciousness.

This all too brief voyage into the minds of birds should, if nothing else, help to banish the expression 'bird brain' into the dustbin of human ignorance. Birds display attributes of higher sentience and special skills just as advanced as those observed in the most advanced of mammals and these have enabled them to succeed in every possible habitat on land and sea. These attributes include tool use, empathy, play, social learning and formal education. Their navigational abilities far exceed anything we can manage. They undoubtedly have gaps in their knowledge but then don't we all? As with all species, their best developed skills and understanding are those linked to the things that matter most to them in terms of the genetic fitness of the species and the wellbeing of the individual. However, they do find time for trivial pursuits. Birds of the air have been given special opportunities through their ability to fly. They have also been presented with special challenges occasioned by their physical fragility and the physical and mental demands of successful breeding. These opportunities and challenges call for a high degree of mental dexterity, more than that required of many mammals who can often get by for the most part on the basis of brute force and ignorance. I am lost in admiration.

Bats

As birds have command of the air in the day, bats have command of the night. They are the only true flying mammals, but adaptation to flight has been remarkably successful since bats account for about 20% of all mammalian species. There are three families of bat. Most species are described as microbats. These are the insect eaters who hunt at night using echolocation. The megabats are largely fruit eaters who operate both night and day. Vampire bats have evolved to live on a diet of blood, which they typically obtain by sneaking up on sleeping cattle. Microbats roost through the daylight hours, hanging upside down in caves, under roofs and other secure spots. Their social structure is similar to that of many mammals that live in groups. Females normally give birth to a single offspring and provide all its support in the form of mother's milk. Males with no parental responsibilities can afford to be polygynous and, like stags, cockerels and elephant seals, establish and defend a harem. Females within the harem form a network of close relationships that include food sharing and, on occasions, suckle each other's offspring.

The hunting behaviour of insectivorous microbats is based on echolocation (aka sonar). The bat emits a series of high-frequency sounds, listens for an echo to detect the location of its prey, and calculates its speed and direction from the Doppler shift: the difference in frequency between the transmitted and received sound. It has a further neat specialisation, an ability to contract the muscles of the middle ear as it emits its squeaks, so as not to deafen itself (26). For long range navigation in the dark, bats can use magnetoreception to locate north and south (at least).

Bats have achieved success and long life despite the high energy cost of hunting for food in the air. Their average life expectancy is considerably greater than that of most land mammals of similar size. Part of this can be attributed to their ability to lower their metabolic rate while at rest through daily periods of torpor during the active hunting season and hibernation during the winter. Their *average* longevity is also enhanced by an infant survival rate considerably higher than that of most bird species. They have achieved this by developing some very advanced social skills, of which perhaps the greatest is the mutually supportive behaviour of the sisterhood.

Animals of the Savannah and Plains

Before human intervention upset the sustainable habitat of the African savannah, the American prairies and the steppes of central Asia, large numbers of herbivores grazed the pastures and browsed the shrubs and trees. Smaller numbers of carnivores killed and ate some of them. On the savannah, the initial beneficiaries of predation may be a pride of lions, followed by a flock of vultures to tidy up afterwards. In the Canadian wild, wolves manage the caribou population. The First Nation people have a wise saying: '*It is the wolf that keeps the caribou strong*'. This is a classic expression of Darwinism. Not only do the wolves control the numbers of caribou but by killing the weakest and slowest they promote the overall fitness of the population. Before the arrival of the Europeans, the predators that sustainably managed the huge population of bison on the American prairies were the indigenous human population.

Although the great majority of the herbivorous species are wild, their numbers are in steep decline because humans are taking away their habitat. Today, most of the animals grazing the open plains are semi-domesticated species such as cattle, sheep and goats. Most semi-domesticated herbivores of today are killed by us, in abattoirs. However, whether wild or domesticated, the environmental challenges to the herbivores free ranging on the open plains and the impact of these challenges on their minds, motivation and behavioural needs are much the same.

Animal Welfare: Understanding Sentient Minds and Why it Matters, First Edition. John Webster.

For about 99% of the span of life on earth, herbivores and carnivores coexisted in a balance of nature that maintained population numbers in both groups. Quick culling of herbivores by carnivores reduced the risk of overgrazing which could otherwise lead to their slow death from starvation. If the carnivores killed too many herbivores, they would go hungry, leading to a fall in their numbers from malnutrition and infertility. Human interference with this balance of nature has affected the animals in many ways, both good and bad. Semi-domesticated sheep and cattle should enjoy a more reliable feed supply, but things can go badly wrong, for example, when too many animals are kept on too little land during conditions of prolonged drought. They will be less likely to experience the daily risk of predation, but the end of their life may be worse as they endure the stresses of handling and transport prior to slaughter.

Environmental Challenges

The big difference between the environmental challenges to domesticated and wild herbivores grazing the open plains is that the former should enjoy some protection from their human herders, the latter face the constant threat of predation. Populations of wild herbivores have adapted to this threat. I repeat: the impala is less at risk of extinction than the cheetah. Here, I think I should briefly reprise my thoughts on the mental processes involved in the survival strategies of herbivores exposed to predation from the big cats on the open plains (Chapter 4) All prey species have a hard-wired programme that sets their *flight distance* from a potential predator. When the predator is outside their flight distance, they will be aware of its presence but see no need to take evasive action. As the predator approaches the flight distance they will, if unrestrained, move away in a controlled fashion in a direction determined by the angle of approach of the predator. If the predator encroaches within the flight distance, they will run off at speed. Survivors learn by experience that they can escape. Incipient threats may be picked up from a distance. Most signals are likely to come from members of the herd who spot the danger first and signal to the others.

For many wild animals that experience nature red in tooth and claw, tasks essential to survival may take up most of their time. It would be a mistake, however, to assume that their minds are exclusively hard wired to these primitive tasks. We can observe wild animals, both youngsters and adults, with time on their hands in the better, environmentally enriched zoos engaging in behaviour that looks to us just like play. I repeat what I wrote in Chapter 2. Some severe animal behaviourists reject the notion that non-human animals engage in play. I cannot accept this argument. Anyone who has watched young lambs playing 'king of the castle' around straw bales would have to be very severe indeed to deny that they were having fun. Young animals, whether lambs or lion cubs have lots of energy and lots of spare time. Most behaviourists would concede that sentient animals are strongly motivated to seek pleasure as a means to feeling good if they do not conflict with the actions necessary for survival. Pleasure may take the form of rest or recreation. Hard working adults may take the chance to rest and luxuriate in the sun, energetic youngsters to play games. One of the reasons that signs of play behaviour are more common in carnivores is because they have more spare time. Otters, for example, who need little time to catch their daily ration of fish, sport on mud slides.

Adult herbivores on sparse pastures are so occupied by the need to take in sufficient food that they have little time or energy for the pursuit of pleasure. However, cows can get satisfaction from time spent comfortably at rest, ruminating in the sun or shade. Groups of young foals regularly enjoy what my wife refers to as 'their mad five minutes': wild races, with their tails held high in the air, typically at sundown.

Animals of the Open Plains

The predominant animals of the savannah and plains are the bovidae, ruminant herbivores with cloven hooves. This family of more than 140 species includes a small number of domesticated species, cattle, water buffalo, sheep and goats. Wild species within this family include bison, African buffalo, antelopes (dik-dik, impala and wildebeest) and musk ox. The males, and, in some species, the females have permanent horns growing out from their skull. Other families include giraffes and the camelidae (camels, llamas, alpacas, guanaco, vicunae), technically classed as pseudo-ruminants. The cervidae, deer, elk (or wapiti), moose and caribou (or reindeer) are also ruminants but differ from the bovidae in having antlers, rather than horns, which develop each year from skin tissue rather than bone and are shed at the end of the breeding season. The Cervidae are principally creatures of the forests so will be considered in more detail in the next chapter. Other animals of the plains include zebras, feral horses and African elephants, all herbivores with the fermentation vat at the hind end of the digestive tract so adapted to take in food slowly by grazing or browsing for long periods.

The apex predators of the plains of Africa are the big cats, lions, leopards, and cheetahs. Apex predators in other regions include the jaguar, cougar and snow leopard. All these animals are obligate carnivores: they have evolved to subsist on a diet of protein and fat, have a limited ability to digest carbohydrates and almost no ability to digest fibre. They must kill or die. Behind the apex predators come the scavengers, such as hyenas. Omnivores like African hunting dogs, or bears (which are creatures of the forests) have a wider range of options but will eat meat when they can.

In the following discussion of the environmental challenges encountered on the open plains and the responses of the different species, it will rarely be necessary to distinguish between wild and domesticated. Inevitably however, more space will be given to semi-domesticated species like sheep and cattle on range since we know so much more about them.

Sheep

Sheep, whether wild or semi-domesticated, occupy environments where they must forage long hours to obtain enough nutrients to maintain good health and support their offspring. They are relatively small, slow-moving animals, with limited ability to defend themselves so depend for their security on strength in numbers. To promote the success of their own family, within a large flock, they need to recognise who's who.

Sheep quickly discover or learn from their mothers to avoid poisonous plants (41). In experimental studies, sheep have shown an aversion to the smell of the wolf, while apparently not disturbed by the smell of bear, which suggests that these aversions are expressions of innate fears, like the monkeys' fear of snakes (58).

Healthy, nutritious plants vary in their proportions of protein and digestible energy (mostly sugars and cellulose). Growing lambs and lactating ewes need a higher ratio of protein to energy relative to adults that require food only for maintenance and adjust their feed selection to provide the protein:energy ratio best suited to their needs (41). The sugar content of high energy grasses is highest late in the day when the sun has done its work, so sheep tend to favour the high protein grasses in the morning and the high energy grasses in the afternoon. Sheep that carry a high burden of intestinal worms select high protein grasses to restore nutrients taken up by the parasites. There is some evidence, not too well established, that sheep with a high worm load will select plants with a high content of tannins having antiparasitic properties.

Sheep have an exceptional ability to recognise individual faces and sounds, especially the sound of their own lambs. Key receptors in the brain pick out the sights and sounds that matter from the background noise coming from the continuous input of unnecessary information. These special skills enable ewes to favour their offspring, and thereby their genetic inheritance within the group. The natural behaviour of ewes is to ensure that their milk goes to their own lambs. They will forcibly reject other lambs that approach them and attempt to feed. Shepherds have developed several methods to get ewes that have lost their lamb to foster lambs that are not their own. A traditional practice was to skin the ewe's own dead lamb and use it to cover the lamb to be fostered. Smothering the foster lamb with placental fluids from the dead lamb may also work just as well. Fostering has become a bigger problem in recent years as a result of selection for fecundity. More ewes are having triplets, which is a problem as they only have two teats. Science, however, has come to the rescue. The hormone oxytocin which facilitates uterine contractions during parturition and milk ejection during suckling is also critical to development of the close emotional bond between mother and child. One can greatly improve the odds of successful fostering in sheep by administration of oxytocin as a nasal spray to the ewe at the same time as you introduce her to a lamb not her own (35). Interpreting this in terms of the simple model of the sentient mind, the oxytocin spray has given positive reinforcement to the arrival of the new lamb by associating it with a strong feeling of pleasure.

The ability of sheep to distinguish and react to sounds that matter is not restricted to potentially hard-wired signals like the call of their own lambs. Christine Nicol filmed a flock of sheep at pasture but accustomed to getting a daily ration of corn. All day, cars, lorries and Land Rovers passed by on the road without attracting a flicker of attention. However, when the specific sound of the farmer's Land Rover was heard in the distance the sheep set off at once in the direction of the feed troughs to welcome his arrival.

In areas where sheep are farmed extensively and at risk from predators, survival of the individual (which is what matters to the individual) has been favoured by the motivation to stay within the flock, especially when under threat. When presented with a potential threat outside their flight distance (e.g. a sheepdog) they act in a measured way. We humans observe sheep to panic when attacked by straying dogs or driven by incompetent herders. However, before we dismiss sheep as stupid because they behave irrationally when stressed beyond the point where they perceive they can cope, we should acknowledge that few of us are able to keep our heads when in a state of panic.

It is possible that the flocking behaviour of sheep has evolved as a result of their semi-domestication. Primitive Soay sheep that lived away from humans and dogs for many generations on the uninhabited isle of St Kilda scattered in all directions when, finally, attempts were made to round them up with dogs. Barely domesticated Blackface sheep thinly spread over wide areas of moorland, e.g. in the Scottish Highlands will stand up for themselves. I once watched a Blackface ewe with a dog in hot pursuit leap a mountain stream, then turn and butt the dog every time it tried to get out of the water. Having skilfully chosen her field of battle, she emerged victorious.

Goats

Goats are similar to sheep in most aspects of form and function. In behavioural terms, they could hardly be more different. While both species are selective feeders, sheep tend to restrict their choice to grasses and herbs growing at ground level, goats are much more inclined to browse small trees and shrubs. In the rainy season when the quality of food available for grazing and browsing is of a similar high quality, goats will forage indiscriminately from both. As the quality of grass deteriorates during the dry season, goats prefer to browse, even climbing trees to get at the juiciest and most nutritious leaves. Goats have a reputation of being indiscriminate and eating anything. In fact, their choice is very soundly based. I know a commercial dairy goat farmer who feeds his lactating nanny goats a basic ration of grass and maize silage and gets far more milk from his animals than the nutrition chemists would predict from analysis of the forage. This is because the goats select the best bits. The leftovers that the goats reject are fed to growing cattle that can thrive on a less nutritious diet. These goats have the luxury of choice, but the motivation of goats to try anything works to their benefit when conditions are harsh. I have observed semi-domesticated sheep and goats in the Sahel region of northern Nigeria at the end of the dry season. The sheep were listless and emaciated. The goats were fit, bright and actively scavenging for anything they could find. This motivation to eat anything in times of need has its downside. When food is scarce, too many goats can destroy the natural habitat.

Goats, unlike sheep, do not rely on the safety of the flock when threatened by predators. They are much more agile than sheep and can make their escape by moving nimbly over difficult terrain. They are particularly well adapted to mountain regions such as the Alps and North American Rockies, where their ability to traverse precipitous rock faces keeps them safe from most predators (*see* Figure 3.1). Because goats are so much better than sheep at looking after themselves, they are much less timid. Chapter 2 introduced the principle that sentient animals promote their survival and wellbeing through an appropriate balance between curiosity and caution. For reasons that are entirely consistent with their own wellbeing, goats operate with confidence at the curiosity end of the spectrum. Goats with time to spare are highly motivated by the pursuit of pleasure, and this includes a strong sense of mischief. I illustrate this with two examples from personal experience. When at the University of Alberta, Canada, I was giving lessons in animal handling to a class of students (mostly ranchers' sons who probably didn't need my advice) in a barn containing several pens occupied by a mixed population of animals. One bored student was giving a highly attentive West African pygmy goat instruction in how to operate the latch that opened the pens. For the next two days, we arrived in the morning to discover that pen was open, and the goats were out. On the third day, all pens were open, and all the animals were out. We were forced to change the latches. Some years later, I was in a barn in Somerset teaching a vocational course on goat farming to an enthusiastic group of middle-class aspirant peasants. A lady with a long scarf and large hat had draped herself elegantly over the side of one of the pens. Being more interested in the behaviour of goats than humans, I watched but did not interfere with one goat who clearly had her eyes on the hat. She sidled up slowly, displaying a series of visual signals to indicate that the hat was the last thing on her mind, then made a sudden successful snatch and ran off at high speed to the far end of the shed. I rest my case.

Cattle

Cattle, like all ruminants, are designed to eat at speed and ruminate at leisure. They have no incisor teeth on their upper jaw. They harvest the grass in a distinctive way, grasping it with their tongue and scything it across the incisors in the lower jaw. This enables them take in food quite rapidly, so long as the grass is long enough to be grasped by the tongue, but puts them at a disadvantage compared with sheep, goats and with horses (which have both upper and lower incisors) who can crop pastures right down to ground level. Cattle are therefore the first to suffer in conditions of overgrazing or drought. On the other hand, grasslands grazed by cattle or, in earlier times, by bison, are inherently more sustainable since these animals are anatomically incapable of destroying them through overgrazing.

Cattle show evidence of food wisdom, especially in relation to minerals such as sodium and phosphorus, which are likely to be deficient on natural grasslands. There is much anecdotal evidence that cattle ranging widely on poor pastures in southern Africa will seek out and chew on bones. This skill has been confirmed by experiment. It appears that they do not recognise phosphorus in the simple mineral form but associate it with the smell of organic compounds coming from the bone (17). Here again, the animal has formed an association between the smell or taste of a food source and how it feels at some time after the meal. These skills appear to be quite restricted. High-yielding dairy cows on highly fertilised spring pastures can suffer and die from acute magnesium deficiency but magnesium-deficient cows are not attracted to magnesium supplements unless heavily disguised in something sweet, like molasses. This lack of magnesium wisdom should not come as a surprise since magnesium deficiency has never been a problem for cattle living a natural life on permanent pastures of mixed grasses.

Wild and semi-domesticated herds of cattle form social groups of females and their calves, usually with one dominant bull. Young bulls tend to live apart from the cows and the dominant bull until such time as one is able to chase the old bull out and move in. Cows establish a stable hierarchy, usually reinforced by signals rather than physical conflict. They do form special attachments within the herd, reinforced by positive gestures such as social grooming. They also engage in play fights, usually involving some harmless pushing and shoving. These encounters may be classified as play rather than aggression because they are often initiated by the subordinate animal.

The dominant bull will regularly scan his harem to identify any cows that are coming into oestrus and will soon be receptive to mating. He recognises the onset of oestrus by sniffing around the vulva and, if interested, demonstrates the 'flehmen' response: he curls back his upper lip and sniffs it. This behaviour helps to transmit the smell to the vomeronasal organ, specifically adapted to recognise pheromones (Chapter 3). This pattern of behaviour is most often seen on the day before the cow is in full oestrus and ready to be mated. The bull makes a mental note that the cow will be receptive tomorrow, returns the next day and mates after little preamble. In commercial dairy herds, where there are lots of cows together, but no bull and breeding is by artificial insemination, the traditional way for us humans to detect oestrus has been to identify the cow that is being mounted by other cows. Cows not in oestrus and therefore not in a high state of sexual arousal, are nevertheless attracted by the smell of a cow on heat; the cow in oestrus is willing to be mounted. In small herds running with a dominant bull, this homosexual display is less common. It does take place, although usually out of sight of the bull.

Cattle are a hiding species. Cows will select a safe spot to give birth at a discrete distance from the herd. After ensuring the calf is alive and well fed, the cow returns to the herd but keeps the calf under continuous observation and will speed to its defence if she perceives it to be under threat. Humans have been killed or maimed by cows because they had accidently come into close contact with a calf hidden in a hedgerow. Cattle in the wild cannot run as fast as horses, but they are big and strong, and they have horns. Consequently, they are more inclined to turn and fight. Young cattle will gang up together and attack dogs. Bulls will defend their harem if necessary. Other animals, including humans, should be able to avoid attack by maintaining a proper flight distance and practising subordinate body language, such as averting direct eye contact.

Wild Bovidae

The wild bovidae range widely in form and function. The antelopes include the tiny dik-dik, the spectacularly athletic impala and the relatively defenceless wildebeest, constantly featured on television in the throes of death from a crocodile or large cat. Nevertheless, the wildebeest exist in by far the greatest numbers, which suggests they must be doing something right. For a number of rather complex reasons, they are about the optimal size to acquire nutrients by rumen digestion. They have the stamina to migrate very long distances to the best pastures at different seasons and the herd memory to know where to find them. Young calves are at the greatest risk of predation by the large cats. They are, however, mature enough to run with their mothers within hours of birth. Moreover, calving is synchronised so that nearly all calves in a herd are born within a few days of one another, which means that in the calving season, the supply of easily available fresh meat to predators vastly exceeds demand. The survival of the wildebeest population depends on strength in numbers.

African buffalo and North American bison are big, strong animals, usually able to look after themselves. Bison have the stamina and skills to migrate successfully over wide ranges. Their social interactions are essentially the same as cattle: bulls with their harem of females, young males staying apart. Wolves are their only significant predator, and they are most likely to kill young calves. When under threat from a pack of wolves, the bison run away in orderly fashion with the cows and calves in the front and the bulls bringing up the rear so that they can turn and attack any wolf that gets too close. Some chases go on for hours. The wolves are most likely to achieve a kill if one or more of the calves becomes separated from the herd. As often as not, the hunt fails.

The musk oxen of the Canadian Arctic have one of the best organised defence strategies. Their predator is also the wolf. Under threat, they form an impenetrable circle. The adults, especially the males, form the outer ring, facing outwards and presenting their formidable horns to the predator. Calves are kept safely in the middle. The only way that wolves can overcome this strategy is by harrying the herd to the point of panic.

Feral Horses

Most so-called wild horses are actually feral. The North American mustang and the Australian brumby are the descendants of animals that escaped to the wild and have adapted very well. The native ponies of the UK (e.g. Dartmoor, Exmoor, Shetland) are described as semi-feral. They are well equipped to survive in harsh conditions but adapt equally well to domestication. Probably the only truly wild horse is Przewalski's, otherwise known as the Mongolian wild horse. They, and the feral horses, do not exist in large herds but form small family groups of perhaps six to 20 animals under the leadership of one alpha mare. Being hind gut fermenters, it is necessary for them to stand and walk to graze for long periods, unlike the ruminants who can consume a large quantity of food in a short time, then lie down to rest and ruminate. Horses have a 'stay apparatus'; an arrangement of suspensory ligaments around the stifle, or knee joint, that enables them to lock it into place so as to rest and even sleep standing up. We think this does not include REM (or dreaming) sleep, so they are motivated to lie down from time to time.

Social intercourse between horses in small family groups involves a lot of body language, position of ears, head and neck, and a range of vocal sounds indicative of dominance, submission or simple curiosity. Other obvious signs include mutual grooming and standing head to tail while swishing their tails to derive mutual protection from flying insects. There can be no doubt that horses form close emotional attachments to favoured individuals. In the domesticated arena of a large stable, this can be with any horse. In the wild, it will almost inevitably be with a relative and therefore favour the genetic success of the group. However, in either case, the conscious motivation of the horse is not to favour its genes but to get the emotional satisfaction that comes from having a friend.

Elephants

There are two main subspecies of elephant, the African and the Asian. Their physiology is similar. Differences in behaviour between the two species may be more a consequence of, rather than a reason for the domestication of the Asian elephant. The elephant is a hind gut fermenting herbivore whose natural habitat is the savannah, a mixture of pasture, shrubs and tall trees. They graze the pastures and browse the trees and shrubs, preferring to browse, where possible. In the dry season, leaves are more nutritious than grass. When food is short, elephants can browse leaves on the highest trees, often by ripping down branches to get at them. In an area overpopulated with elephants, the trees and shrubs suffer the most damage and the mixed vegetation of the savannah degrades to open grassland.

The social and sexual behaviour of elephants is similar to that of the other large herbivores. They form small family groups of about ten animals, led by an alpha female. Males tend to stay outside the groups unless there are females in oestrus. Male elephants use exactly the same procedure as cattle and horses to identify females in oestrus, namely sniffing around and using the flehmen response to recognise the sex pheromones. Homosexual mounting behaviour is observed in both males and females. Stable social intercourse within and between family groups involves some ritualised aggression, e.g. spreading the ears so as to look bigger, to establish the hierarchy. Bonds of friendships are established and maintained by mutual grooming and play behaviour. Young elephants engage in play and get adults to join in. The mutual play of young elephants cannot be interpreted simply as training in survival skills like pulling down trees.

The biggest and most interesting difference between elephants and the other large herbivores concerns the extent to which they display evidence of the fourth circle of sentience, namely the capacity for mental formulation. This capacity must, as in all sentient animals, be based on feelings and thought (aka emotion and cognition). A great deal has been written about the emotional depth and cognitive capacity of elephants, based on experimental studies and observations of their natural behaviour. They display signs of distress and a sense of loss following the death of a member of the group, especially the group leader and would appear to mourn over the bones of a fallen comrade (18). There are several reports of elephants going to the assistance of an injured member of the family group. It would be stretching reverse anthropomorphism too far to try to imagine just how they feel at times like these. However, I believe that it is entirely valid to conclude that they do experience strong, lasting emotional feelings for others and that these are influenced by past mental formulations: memories of good times together, loss of companionship and possibly concerns about the future.

Experimental evidence as to their cognitive abilities is somewhat inconclusive. They use simple tools, e.g. break off leafy branches for use as fly swatters, in this respect demonstrating cognitive abilities similar to primates. This is intelligent, although not especially profound and probably learnt by observation from others. When assessed for intelligence using the classic experimental approach of presenting them with an increasingly complex sequence of tasks in order to solve and receive a food award, elephants manage some of the simpler tasks quite well, although no better than rats, far less well than corvids or, indeed, squirrels.

Evidence as to the extent that elephants demonstrate properties of the inner circle of sentience, namely conscious self-awareness, is equivocal. Mirror tests (p. 61) with elephants have generally been inconclusive. Moreover, the demonstration that an animal can pass the mirror test falls some way short of proof of self-awareness linked to the concept of theory of mind, the capacity to put oneself into the mind of another and imagine 'how would they feel, or what would they do, if I act in this way?' Despite the lack of experimental evidence for theory of mind in elephants, I believe that the concern they show for other members of their family group strongly suggests that they have the capacity to acknowledge the feelings of others, although, because they are likely to be members of the family, this does not necessarily equate to true altruism. On balance however, I think it fair to conclude that these forms of behaviour may be manifestations of the deepest circle of sentience, namely full consciousness.

Estimates of the cognitive ability of elephants, based on experiments designed by us to test how well they compare with us in things we can understand, lead to a familiar conclusion: they can perform a series of simple tasks about as well as a small child. Here again, however, these tests fail to address skills we cannot begin to comprehend, the most obvious being spatial memory and navigational skills. Families will travel long distances to reach sources of food and water. They may sometimes be guided by the sense of smell, for example, to sources of salt. In this regard, they are similar to cattle and goats. However, the most realistic explanation for their behaviour in relation to the search for food and water is that they are led by the alpha female who knows exactly where to go. Migrating elephants may get a general sense of direction from the position of the sun and this, of course, implies that they, like pigeons, can tell the time of day. It is, however, almost impossible to avoid the conclusion that the lead elephant has retained in her mind the map of a largely featureless terrain, extending over hundreds of square kilometres. As with the birds, these map-making skills far exceed our human capacities.

There can be no doubt that the elephant is a highly sentient, highly intelligent animal with a strong commitment to family. This begs the question 'How did this come about'? The environmental challenges presented to the elephant are similar to, and in many instances, less threatening than those experienced by other large grazing herbivores so, in this sense, there has been no special pressure to get smart. One traditional argument has been that elephants have a large brain size, with a well-developed cerebral cortex, rather similar to that of dolphins and humans. Here again, the brain size argument is that it cannot account for the corvid paradox: the brain of the crow is tiny compared with that of the chimpanzee, yet they are every bit as intelligent. An alternative, no less plausible explanation is that 'the elephant's child' (36), by virtue of the long time it takes to reach maturity, (combined with a '*satiable curtiosity*') can experience the benefits of a prolonged education. It is a biological fact of life, at least among homeotherms, that the larger the animal, the longer it takes to mature. Some years ago, James Kirkwood and I explored the relationship between mature size and time taken to mature in a selection of mammals and birds (37). In all mammalian species, with the exception of the primates, there was an almost exact correlation between mature size and time taken to mature. Precocial birds, those able to forage for themselves from the time of hatching, showed a similar pattern but maturation rate was faster than in mammals. Altricial birds, restricted to their nests and fed by their parents to the point of maturity, grew faster still. We argued that, in the interests of survival, all these animals proceed to independent maturity as fast as possible within limits set by the need to function properly at all stages of growth. Altricial birds, subject to few physiological demands during growth beyond those of digestion and metabolism, were the quickest to attain maturity. In later work, James extended this analysis to include the primates and demonstrated a progressive slowing down of maturation rate relative to body size, beginning with the new world monkeys, through the great apes to humans (38). He argued that in primate species, including humans, the survival advantages of improved understanding as a consequence of prolonged education under the protection of parents and other concerned adults became greater than the need to acquire the physical benefits of maturity as soon as possible. Maturation rate in the elephant does not depart from the standard line for all non-primate mammals but by virtue of being very large, the elephant has a prolonged childhood and so is able to benefit from a prolonged education. This suggests that selection for intelligence has not been a major factor driving the evolution of the elephant. They are, however, able to profit from the benefits of a good education within a stable and supportive family group.

Predators

The apex predators of the African savannah are the large cats, lions, leopards, and cheetahs, with the lion at the top of the pyramid. These are all obligate carnivores, i.e. they have to kill in order to live. Other predators and scavengers include hyenas and African hunting dogs. Most of these animals hunt in small family groups to increase the chances of success but they are in direct competition with other groups in the pursuit of food and they do not depend for their security on safety in numbers. Conspecifics outside the immediate family are therefore viewed as competitors.

Lions are the most social of the large cats. A pride may consist of up to 20 animals: lionesses and their cubs plus up to four males. Lionesses do all the hunting, A single lion can bring down a zebra, but it takes more to tackle a buffalo or giraffe. Unlike wolves of the northern forests that may pursue their prey for hours, the large cats of the tropical plains stalk their prey to get as close as possible then attempt to capture it in a short sprint. If they don't manage a quick kill, they give up. This is adaptive behaviour in the sense that it conserves energy. Lions, and all cats, spend many hours asleep. There is another subtle feature of anatomy that favours the escape of the ruminant herbivores that are their prey. Ruminants, but not cats, have a built-in heat exchanger that enables them to maintain a cool head while making their escape. The arterial blood supply to the brain passes through the carotid rete, a complex delta of small vessels where the arteries are in direct proximity with veins carrying blood from the nasal cavity that has been cooled by the passage of air as it passes over turbinate bones (heat exchange plates) in the nose. During escape, when both metabolic rate and respiration rate are increased, deep body temperature rises but brain temperature remains constant. Some big cats may choose to give up the chase to conserve energy: they decide it is not worth the effort. However, many are forced to give up because of mental confusion associated with incipient hyperthermia.

Lions will steal food from smaller leopards and cheetahs. Leopards seek to protect their catch by dragging them up trees. All the apex predators, including lions, can be harried off their kill by a pack of hyenas or African hunting dogs. Mature lion males will also kill cubs not fathered by them and then mate with their mother. This is a particularly brutal demonstration of the genetic imperative. Predators that become unable to hunt as a result of injury or age, grow progressively weaker and die of starvation or, more likely, succumb to attack by a pack of hyenas. There is little evidence of a social contract among carnivores. Most of their lives are antisocial and brutish.

Animals of the Forest

The most extensive areas of natural forest are the tropical rainforests of the Amazon, Congo and Indonesia and the boreal forests, or Taiga that extend across the northern regions of Asia, Europe and North America. The tropical rainforests are often described as jungles, but this word has gone out of favour among academics. Because they are very warm and very wet for most of the year, they support a very high density and very high complexity of plant and animal life in three dimensions, the vertical dimension extending from underground burrows to the top of the tree canopy.

Plant and animal life in the boreal forests is restricted by poor soil and long, freezing winters. The number of species that can thrive in this habitat is low and can be considered within three categories: herbivores, like deer and caribou that can exist on leaves and lichen; their predators, such as wolves; and animals ranging from ground squirrels to the grizzly bear who have adapted to the stress of winter by retreating into shelters and going into hibernation.

The Boreal Forest

As a habitat for animals, the Taiga, or boreal forests are best considered together with the open tundra. The main flora of the forests are the coniferous trees, pine, spruce and larch, the leaves of which have little nutritive value, even for ruminants. A few deciduous

Animal Welfare: Understanding Sentient Minds and Why it Matters, First Edition. John Webster.

trees, willows and birches provide digestible leaves in season. Ground cover below conifers is sparse: a few shrubs and grasses, mosses and lichens. In the summer, when the snow melts, large areas of the forest are under water and this aquatic environment supports the growth of highly nutritious vegetation. At the northern limit of the boreal forest, the trees become progressively sparse and peter out altogether at the tree line, north of which is the tundra, the zone of permafrost that, during the short summer when the snow recedes, reveals the muskeg, a water-saturated peat bog held together by sphagnum moss and supporting the growth of some dwarf shrubs, sedges, grasses and lichens. The sodden summer environment of the forest and tundra also supports a prodigious quantity of blood sucking insects such as mosquitos and ticks. The slightly less severe southern regions of the boreal forest contain a higher proportion of deciduous trees, shrubs, aquatic plants and fruiting plants that favour bears, beavers and some of the more static Cervidae like the moose.

Cervidae

Most of the large herbivores of the boreal forest and tundra are the Cervidae. These are given different names in Eurasia and North America, and this can cause confusion. The Eurasian, semi-domesticated reindeer is the same species as the wild North American caribou. A moose in North America is an elk in Europe. What the Canadians call an elk, or wapiti, is a large subspecies of what Scots would call a red deer.

The largest population of the Cervidae are the caribou, or reindeer. Taxonomists identify many subspecies. For our purposes, they may simply be classified into three groups: the migratory caribou, the settled forest caribou and the semi-domesticated reindeer. The migratory caribou of North America spend the frozen winter months in the forest, where the cold and wind are less severe, browsing dead and dying leaves from the few deciduous trees and shrubs, and foraging ('rustling') under the snow for digestible lichens and mosses. Like most other ruminants, they exhibit nutritional wisdom. Lichens are higher in energy and mosses higher in protein. Pregnant caribou that need more protein to sustain the developing embryo, select a higher ratio of mosses to lichens. As the thaw sets in, they move out onto the tundra and gather in enormous groups, ranging in numbers from 50 000 to 500 000. During summer, they migrate across the new growth of vegetation on the tundra, travelling 20–50 km/day, perhaps 5000 km over the entire season at speeds up to 60–80km/hour. Calves born during the migration are up and running within minutes of birth. But why run so far and so fast? The conventional explanation is that the density of vegetation in the tundra is sparse and therefore they need to forage over a wide area. However, they could save a lot of energy by simply ambling forwards over the muskeg, grazing as they went. The strong inelastic motivation to run must be driven by a powerful, presumably aversive stimulus. It is difficult to escape the conclusion that the main reason they run, and run is in a vain attempt to escape the plague of external parasites, including flies and blood-sucking insects such as mosquitoes. The behaviour of fly ridden caribou is similar to that seen in cattle when the warble fly (oestrus ovis), otherwise known as the gad fly, was endemic in the UK. This behaviour has always fascinated me as an expression of mental state. The warble fly has a distinctive sound, but it does not cause any physical discomfort to the animal at the time. Its sole purpose is to lay eggs discretely on the hair of the legs. Pain comes much later, when the egg becomes a large larva that migrates through the muscles of the animal to emerge months later through the skin of the back. The panic reaction ('gadding') of ruminants like cattle and caribou to these insects appears to be a hard-wired response to a sound stimulus associated with a painful sensation that will occur well into the future.

One subspecies, the boreal forest caribou, is said to be sedentary, a strange description of an animal that spends most of its life on its feet. These animals live in the forest throughout the year where the microclimate is better, but food is scarcer. The forest caribou is recognised as an endangered species, partly as a result of loss of habitat. However, population numbers were always low compared to the caribou that spend the summer months on the tundra. The migrating caribou is undoubtedly the more successful subspecies but at a much greater cost to the individual. Predators include the wolf packs who have to run long and hard to get a kill, and golden eagles that can snatch a young calf. However, the main natural killers of the caribou are parasites, both internal and external. Heavily parasitized animals become weaker and less able to escape the wolf pack. While the wolves may have struck the final blows, the parasites were the agents of their destruction. In strictly Darwinian terms, the migrating caribou must be classified as a highly successful species, but it is a strategy that carries a high cost to the individuals. One cannot attribute their success to a highly developed mind.

The reindeer of northern Eurasia are the same species as the North American caribou. They have been domesticated for at least 3000 years, initially as a source of meat and hides. Habitat and diet are essentially the same in the Eurasian and American Arctic, the difference being that in Eurasia, migration between summer and winter grazing sites is managed by the reindeer herders. For the most part, the animals are herded *en masse* and given little or no personal attention, yet individual reindeer can be trained to be ridden, carry loads, pull sleighs and entertain children as Santa's helpers. Selection over millennia for genetic traits favouring domestication will have led to a divergence in personality between the amiable reindeer and the wild caribou but this does not address the question as to how this domestication began. One possibility is that the human population was higher in the Eurasian arctic so that humans rather than wolves became the apex predator. In these circumstances, it would have favoured both species to evolve a form of coexistence based on herding rather than hunting.

Other Cervidae, the elk (north America) or red deer (Eurasia) and moose (north America) or elk (Eurasia) are natural inhabitants of the forest, populating the less severe climates of the southern boreal region and other areas of temperate rainforest where the tree cover protects them from the worst of the weather. Moose are solitary creatures, favouring the marshy forest areas with highly digestible aquatic plants. Adult moose are big enough to deter most predators. The major scourge in their watery habitat comes, once again, from the blood sucking insects, especially ticks. In Scotland, red deer are viewed as monarchs of the treeless highlands, but the choice has not been theirs but ours, the hunters and stalkers. Red deer are more sensitive to cold than sheep or Highland cattle, more inclined to seek shelter, and more likely to die in late winter, not from acute hypothermia but because their limited energy reserves of body fat become exhausted. Their social structure is typically made up of small family groups of hinds and calves, with the males remaining separate for most of the year in bachelor groups.

In the breeding season, males will compete with one another to establish the largest possible harem they can sustain within a breeding territory called a lek. Females, not yet in oestrus, choose to join the lek not at this stage through sexual attraction to the dominant stag but to seek his protection from harassment by all the others. Red deer rarely form relationships outside the family. Many years ago, the UK Farm Animal Welfare Council was debating welfare problems involved in the transportation and slaughter of farmed red deer. Ruth Harrison, the compassionate but unsentimental mother of the farm animal welfare movement suggested that it would be more humane if deer were shot in the field. As always, she based this on personal observation of individual deer being shot while eating with others at a feed trough. The shot deer fell to the ground. The others carried on eating. Her suggestion was rejected on grounds of health and safety, but it was very revealing of the deer's state of mind. It would not, I think, have been the same with elephants.

Beavers

The popular image of the beaver is that of an animal whose habit of gnawing wood exceeds the demands of due diligence and borders on the mindlessly obsessive. Needless to say, this image is incorrect. Beavers are large members of the rodent family. Like capybara, they are strict herbivores, hind gut fermenters, with most of the fermentation taking place in the large caecum, set aside from the mainstream of the digestive tract, to prolong the retention of fibrous food and allow sufficient time for microbial fermentation. The principal food source for beavers is lignocellulose. They obtain this from eating bark and the underlying cambium, mostly from young deciduous trees like aspen and poplar but also willow and birch, however, beavers obtain no significant nutrients from the main structural elements of tree trunks. They will also eat grasses and roots. Contrary to popular belief, beavers do not eat fish.

The most conspicuous and impressive demonstration of intelligent behaviour in beavers is their ability to build and maintain secure, weather protected lodges in which to store food for the winter and raise their offspring. These are constructed on sound architectural principles of wattle and daub. First, they assemble a structural framework of strong vertical poles, build up the walls and roof with a criss-cross lattice of smaller branches, then finally seal and insulate walls and roof with mud and moss. When freeze-up occurs, the exposed area above the water seals rock hard and acquires a thick covering of snow. This provides excellent insulation for the accommodation and acts as an impenetrable barrier to predators like wolves and coyotes. Separate dams are constructed both up- and downstream to regulate the height of the water so ensure safe entry to the lodge underwater and dry, comfortable accommodation above (Figure 9.1). A typical lodge has two 'rooms', a drying-off area close to the entrance and a separate, elevated, dry sleeping area. Building a new lodge and dam can take all summer. Once the family have established a well-made lodge and dams to control water level, they may not carry out further maintenance work until the approach of the first frosts. Much of the summer may be spent roaming in the forests obtaining sustenance from the bark of young deciduous trees and shrubs. They will sometimes fell young trees in the spring and store them until needed in the fall. They may even spend time building and

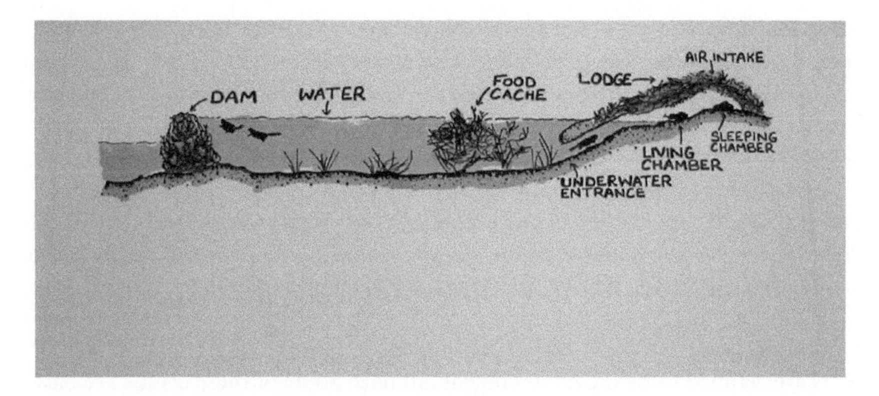

Figure 9.1 A beaver dam and lodge. Note the separate 'drying-off' and sleeping areas.

extending service canals to enable them to float logs down to the lodge rather than lug them overland.

This has to be one of the most impressive demonstrations of manufacturing skills combined with strategic planning within the animal kingdom. Much of this intelligence will be hard wired, but the skills are developed by learning and experience. Beavers are monogamous and both parents contribute to the rearing and education of their offspring. Most young beavers stay within the family for 2–3 years. In years one and two, they spend much of the time playing. In year three, they spend more time assisting their parents with construction work and helping to rear the youngest kits by grooming them and bringing them food. Beavers are a classic example of the extent to which intelligence and mental development are dependent on a sustained education in a stable family setting.

Bears

The brown, black and grizzly bears of the boreal forests are members of the family Ursidae taxonomically classified within the order carnivora and are often described as doglike carnivores. This is an unhelpful definition. These bears are opportunist omnivores, eating anything they can digest. In verdant, mixed forests, about 80% of the diet of black and brown bears will be of vegetable origin: highly digestible shoots and leaves in the spring, fruits in the fall, honey when available. Their highly sensitive sense of smell enables them to locate sources of food underground such as roots, and caches of nuts hidden by squirrels. They will also consume as much food of animal origin as possible. This includes salmon during the salmon run. This requires skills that young bears must acquire by observation and trial and error. They also kill young deer. In Labrador, where vegetation is scarce, they are more dependent on killing young caribou, polar bears have little option but to kill to eat, although they too will scavenge anything from garbage dumps.

The survival strategy of forest bears is to eat and lay down as much fat as possible, then retire to the comfort of a winter den and enter into partial hibernation (see p. 47). Females usually give birth every second year. Bears without cubs mate in May but implantation does not take place until November. Cubs are born in February, very immature and weighing less than 400 g. The mother feeds them in her den until they emerge in the Spring. For the first summer, the mother is fiercely protective of her cubs. This requires a solitary life away from threats, especially from adult male bears.

The expression 'doglike carnivores' presents a distorted image of the social life of bears, which is very different from that of wild dogs or wolves. Bears are solitary and antisocial. They will gather at a plentiful food source, like a salmon run or a rubbish tip, but this will involve a lot of aggression. They do share a liking for play with carnivores, especially, although not exclusively, as youngsters. These games take the form of mock fights, chases on the ground and up trees, and sexual behaviours such as mounting. In the wild, these games are kept within the immediate family.

Bears are claimed to be highly intelligent animals. Some of these claims are based on evidence that I have already rated as suspect. They have large, convoluted brains, which puts them on a par with elephants and primates, but they have not been shown to possess the problem-solving skills of a crow. I concede that crows are easier to work with in the laboratory. Circus bears can be trained to do tricks devised by us. Interpreting this as a measure of high intelligence is an expression of the anthropocentric fallacy. However, manual dexterity is one of the special skills of bears and is put to use in such tasks as opening garbage cans when living in too close proximity to human habitation. A more convincing case for the claim of high intelligence is that they share with other smart species, such as rats and elephants, two of the most important experiences likely to develop the mind. Like rats, they live in a complex and changing environment that calls for learned responses. Like elephants, they live with their mothers for an extended period and so benefit from the blessings of a prolonged education.

The Tropical Rain Forests

The most striking feature of the hot, wet tropical rainforests is the canopy of trees, many reaching heights of 30–45 metres as they compete to get energy from the sun for photosynthesis. In the densest rainforests, less than 5% of sunlight will penetrate to the forest floor. The Amazon rainforest contains approximately 16 000 species of tree and 2.5 million species of insect. Even if we ignore the insects, it would be quite unrealistic to attempt an inclusive review of sentient animal life in the rain forests. Animals of the Amazon basin that can lay claim to at least one property of deep sentience, namely perception, include 1294 species of birds, 427 species of mammals (including 160 species of bat), 428 species of amphibian and 378 species of reptile. Less dense forests, with less rainfall, can sustain larger populations of ground-living animals like deer, pigs and other smaller mammals, and these attract predators like the panther, leopard and tiger.

Most of the vegetable and animal life of the rainforest in contained within the complex, three-dimensional structure of the canopy. The thousands of different species of mammals, birds, amphibians and reptiles that occupy the Amazon rainforest have established and sustained themselves in a myriad of mini ecosystems of plants, herbivores and carnivores adapted to particular environmental niches, engaged in their own sustainable struggles for existence but seldom in direct conflict with other ecosystems (other than through human interference).

It would be quite unrealistic to attempt a wide-ranging review of the effects of life in the rain forests on the development of the sentient mind. Birds and bats have been discussed already. Here, I restrict myself almost entirely to consideration of the primates ranging from the New World monkeys of the Amazon rain forest to the great apes of the Congo basin and the Orangutans of Indonesia. But first, a few words on snakes.

Snakes

The behaviour of snakes has attracted more attention than that of most reptiles, possibly because some people find them particularly fascinating, and many keep them as 'pets'. Snakes of the tropical rain forests include the boa constrictors of South America and Caribbean that kill their prey by crushing. As described in Chapter 3, the special senses used by snakes to hunt their prey are mostly those of smell and taste. They flick out their tongues to sense the air and transfer the information to the vomeronasal organ for interpretation. This gives them detailed information not only as to the presence and size of potential prey, but also its location and velocity (i.e. speed and direction) of movement. For snakes that hunt at night, this conveys more precise spatial information than they can get from vision. The Cuban tree boa attaches its hind end to a branch and stretches its head out to locate fruit bats as they congregate around the opening of their caves at night. It doesn't succeed with every strike, but it gets enough food to live on. Snakes will improve their hunting skills through constant practice, but this is not something that requires higher thought. It is a polished reflex akin to that of a high-class slip fielder at cricket. There is, however, some evidence of advanced behaviour in tree boas, namely coordinated hunting. Rather than select an isolated lair to avoid competition with other boas, they increase their chances of a successful kill by congregating in numbers at the cave mouth, thereby creating a palisade of snake heads, which the bats for all their skills in flying and echo location, find hard to traverse.

The king cobra of S.E. Asia has acquired a reputation for intelligence, largely through folk lore. They can apparently recognise their 'owners', individual snake charmers, and have enough spatial sense to define, remember and defend their own territories. I don't regard this as particularly worthy of note. There is, however, one set of studies of snake intelligence that is highly relevant to my theme because it tells us more about behavioural scientists than about the snakes themselves. This involves the classic test of cognitive ability by measuring their ability to find their way through relatively simple mazes in order to achieve a reward. As described in Chapter 3, rats can not only discover and remember the route to the right exit but also display latent learning; the ability to store information of no immediate use, such as pathways through a maze when there is no incentive to choose. Scientists have carried out many maze running experiments with snakes and, in general, were not impressed by their apparent cognitive abilities. The main reason for this was that the snakes did not find the questions to be particularly interesting. In a much more appropriate series of experiments, snakes were placed in an open enclosure, likely to be aversive, with a number of exit holes, only one of which was open. In repeat trials, snakes learnt the way out (as measured by the time to exit) as quickly as rats. They were highly motivated to solve the problem to meet their need for security, something that really mattered. The moral of this tale is that you can only get a proper idea of the extent of intelligence in animals if you ask them the right questions.

Primates

Animals of the canopy must be able to move with ease and in safety both within the trees and on the ground. Most primates have developed extreme agility. All have fingers and thumbs that enable them to grasp and hold on to branches but also enable them to manipulate food sources and use tools to get at foods. Some monkeys have powerful tails that serve as excellent climbing aids. The evolution and behaviour of the primates provides a good example of the resolution of competition between different species for resources through adaptation to different environmental niches, e.g. chimpanzees live and feed mostly within the canopy, the heavy vegetarian gorillas more often at ground level. This does not, however, resolve the problem of competition for resources within species living in close proximity and this inevitably creates the potential for conflict. Adaptation of social species like the chimpanzee to this threat leads to behaviour than can best be described as tribalism. The animals develop close bonds within their society, and these extend beyond that of the individual family to that of the group, or tribe. This is coupled to an aggressive attitude to neighbouring tribes. Orangutans, who spend nearly all their lives within the canopy, are solitary, except for mothers and their young. This may be a consequence of the fact that, unlike the African primates, they have experienced little competition for resources (except from humans).

Colonies of primates are sustained and enriched by many expressions of deep sentience. Social strategies include complex patterns of visual and verbal communication, and emotional expressions of friendship between tribal members (41). The most convincing evidence that this behaviour is not merely instinctive but reflects local cultural traditions is exemplified by the extreme difference in social and sexual behaviour between chimpanzees and bonobos, two genetically near-identical great apes in near-identical environments but living apart on opposite sides of the Congo river (9). Chimpanzees have evolved into a patriarchal society, with alpha males protecting the tribe from aggressive neighbours and maintaining strict discipline within. Bonobos are a matriarchal society, displaying little aggression and getting much of their emotional release from sex. Both male and female bonobos solicit and engage in regular sex, whether females are in oestrus or not. Despite the penchant of chimpanzees for aggression and bonobos for promiscuity, females in both groups give birth at about four-year intervals. The two divergent cultures are equally successful.

One of the most striking similarities between the primates, especially the great apes, and humans is in the anatomy and physiology of the digestive tract. Approximately 85% of the diet of orangutans, chimpanzees and bonobos is made up of fruits that provide a rich source of energy in the form of sugar but little protein. Their need for protein usually involves the consumption of food of animal origin from any accessible source. This will include insects, lizards, small mammals including monkeys and, in the case of chimpanzees, occasional cannibalism. Ground-dwelling gorillas have less access to fruit and are much more dependent on a diet of leaves containing cellulose which provides nutrient energy and some microbial protein after fermentation in a greatly expanded hind gut, which is why they have large bellies and less motivation to kill animals for meat. The fruit-eating primates of the canopy share a limited capacity to digest uncooked starch with humans. In a strictly physiological sense, both they and we may be described as fruitists. Before the birth of agriculture, humans, like chimpanzees, were hunter/gatherers. Agricultural communities developed as humans learnt not only to cultivate starchy cereal crops but also how to grind and cook them to ensure that they could be properly digested. Vegans of today are engaged in an entirely honourable practice, but one made possible by human culture, not primate physiology.

The high energy, highly digestible, highly available, nature of the diet of most primates means that they need relatively little time to search for food and comparatively little time for rest and digestion after the meal. This gives them much more time for play. The claim that play is no more than learning behaviour is blown out of the water by some of the more gleeful aspects of play in primates. Here is a case where nature programmes on television can teach us a lot. One sublime and entirely convincing sequence has been that of young macaque monkeys behaving like young children, performing ever more spectacular leaps into their favourite swimming pool, swimming to the side and rushing round to do it again.

Throughout this book, I have steered clear of the anthropocentric fallacy that the more human-like the animal, the greater its emotional depth, its intelligence and the more worthy it is of our respect. At this point however, I have to concede that the primates do share a lot of human traits. These include our love of play, but also our propensity for aggressive tribalism. This does not necessarily make them any 'better' than other species of sentient animals. They are just more like us.

Close Neighbours 10

This section so far has explored how the minds of sentient animals have been shaped by the environments inhabited by themselves and their ancestors. We have seen how, over the generations, they have selected the tools best served to their special needs and, with benefit of the properties of deep sentience, honed their skills and developed their minds to promote the wellbeing of themselves, their descendants and, within social species, the wellbeing of the tribe. This chapter considers the impact of domestication on the development of sentient minds by examining our close neighbours: animals for whom the greatest environmental challenge has been the impact of close contact with humans. These include our companion animals, dogs, cats and horses, whom we regard and treat as individuals and the farmyard animals, pigs, chickens and dairy cows, who may or may not attract our affection but who are, for the most part, regarded *en masse*. Both have had to adapt to a life in which they are dependent on us for their maintenance and wellbeing. I also include within the category of close neighbours, wild rats, urban foxes and other species who have had to adjust their lives to the impact of civilization.

We humans have domesticated animals to serve our own wants and needs, for food, for work, for sport, for love and companionship and, with the best will in the world, our approach to them has been determined by our needs, not theirs, so inevitably reflects our attitudes to them, not theirs to us. We may choose to make a pet of a pig or a rat, rather than a dog or a cat, but that only adjusts our approach to one individual,

Animal Welfare: Understanding Sentient Minds and Why it Matters, First Edition. John Webster.

not the entire species, however much we might wish otherwise. There was a rather touching illustration of this in a TV commercial some years ago. A cattle rancher from the southern USA has just won the lottery. He drives his enormous automobile out to see his herd of cows, waves his winning cheque and announces: 'Good news ladies, you have all been reclassified as pets!'. If only.

History of Domestication

In keeping with my central theme, this section will consider the history of domestication through the eyes of the animals that have become domesticated. I use the word domesticated throughout, rather than domestic because it describes a process. In taxonomic terms, there is no such thing as a domestic animal. Nevertheless, some species have adapted better to domestication than others. According to archaeological evidence, the first species to become domesticated was the wolf, about 15 000 years ago, before the dawn of agriculture when humans were still hunter-gatherers. Some wolves began to scavenge for food around the fringes of human encampments. Those who did found rewards that contributed to their fitness, the numbers of this sub-population increased, their minds and bodies evolved and, in time, they became what we now call dogs. In terms of both fitness and wellbeing, this process of domestication carried significant benefits to the individual animals that chose to become domesticated. In time, humans learnt to derive benefit from employing dogs as guardians and aids to the hunting of wild animals. However, I guess they threw out scraps to these semi-domesticated wolves just as much because they liked to have them around. Today, dogs make wide-ranging contributions to society. These include herding sheep, guiding the blind, sniffing out bombs, drugs and diseases and comforting the afflicted. Most people however choose to live with dogs simply because they add to our quality of life.

The next wave of domestication came with the birth of agriculture, about 12 000 years ago. Humans who lived in areas suitable for the growing of crops were able to form larger, more settled communities. Others, on more marginal lands, adopted a nomadic lifestyle. However, both gained benefits from the management of a small number of herbivorous species, like sheep, that were relatively easy to semi-domesticate and provided an excellent source of food and clothing. These largely defenceless herbivores derived significant benefits from controlled feeding through movement to good pastures and protection from predators, often employing already domesticated dogs that happened to be to hand.

The domestication of cattle came later. It is not clear how humans first managed to domesticate the large and aggressive auroch but cattle and similar species, such as the water buffalo and yak, have proved of immense service to humankind for millennia as providers of work, milk, clothing, meat, fertiliser and fuel. Mass consumption of cattle meat as beef is a very recent development, made possible largely by refrigeration. Throughout most of the history of agriculture, cattle were more valuable alive than dead. The house cow provided a continuous supply of nutritious milk and probably an income from milk sales. A small number of male cattle would be kept to pull the plough. Most males failed to outlive special occasions that called for the slaughter of the fatted calf while it was still drinking from its mother.

It is significant, however, that only a very small proportion of mammalian species have successfully adapted to domestication. In the case of birds, the proportion is tiny. Only chickens, turkeys and other largely flightless birds have benefited, in a strictly Darwinian sense, from an existence entirely determined by humans. In terms of simple numbers, the broiler chicken can be described as the most successful vertebrate species on earth. This is an extreme example of the truth that Darwinian fitness of the species does not necessarily equate with wellbeing of the individual.

The fact that domestication has only succeeded in very few species suggests that it only works on the rare occasions when it carries some benefits to both parties. The process has features in common with Thomas Hobbes' concept of the Social Contract in human society whereby individuals and groups sacrifice certain freedoms in order to obtain social benefits and thereby avoid lives that are 'solitary, poor, nasty, brutish and short'. There are obvious limitations to this analogy. Whatever the quality of life may be for animals reared for meat, their lives will be short. The social contract, as applied to farm animals, may be summarised in the bleak but honest phrase: '*for the first six months, the farmer feeds the pig; for the next six months, the pig feeds the farmer*'. For a true social contract to exist, all parties must relinquish certain freedoms but retain certain rights. This means that all parties should be involved in the negotiations, and, for obvious reasons, the animals are unable to negotiate terms. Any form of social contract will be corrupted when one party gains absolute power without natural or legal checks and balances. While I do not accept that the exploitation of animals *en masse* in the name of food production necessarily abuses the principle of justice with respect to sentient farmed animals and the living environment, abuses do occur, especially within some forms of large-scale factory farming, driven by machines and fossil fuels and insulated from the needs of the land and the life of the land. In all these matters, we carry total responsibility to ensure justice. We are the moral agents; the animals and the land are the moral patient. I shall have more to say on this in the final chapter.

Before we conclude that domestication is always unjust because we humans always hold the cards, consider the case of the so-called domesticated cat. Rudyard Kipling's quintessential 'Cat that walked by itself' (36) chose to enter a human household entirely out of self-interest, offering little in return. In agricultural societies, cats can be useful predators, keeping the grain stores free of rats and mice, but even this is done entirely on their terms. In the home, cats adapt well to being the recipients of our love. It is difficult to argue with the old saying that 'dogs have owners, cats have staff'.

Artificial Selection and Unnatural Breeding

One of the most significant impacts of domestication on the animals is that we take control of their reproduction. In the wild, breeding is by natural selection. They choose their mates, those that make the best choices have greater reproductive success, their offspring carry a higher proportion of genes consistent with fitness in their natural habitat, and so the process of adaptation to the environment through natural selection proceeds to promote the survival of the fittest. This is not a planned operation but, viewed in hindsight, it is, by definition, the one that works, barring catastrophic change to climate and/or habitat.

When we manage animals in harsh and challenging environments such as the Highlands of Scotland or the Siberian plains, we adopt a breeding policy that is close to natural selection so as to maintain the survival of the breeding flock or herd. An 'improved' ram or bull may be used on a selected number of females to produce meatier lambs or calves to be reared on a higher plane of nutrition for early slaughter. Modern, highly industrialised agriculture has removed the challenges of the natural environment by confining animals in climate-controlled barns and delivering their food by machines. In many cases, this intensification has been supported by the widespread use of antibiotics to control bacterial diseases linked to high-density living. When we remove the natural challenges that have defined the natural form and functions of animals, we can practice artificial selection for traits that best serve our interests, like more milk and more meat. Measured strictly in terms of productivity, the success of the agricultural industry has been spectacular. Chickens are now raised to slaughter weight in less than 40 days. Cows that once produced ten litres of milk per day can now produce sixty. All this has happened in the last 60 years. The genetic potential for these enormous changes was there but could only be expressed when we changed the rules. It should come as no surprise to learn that extreme selection pressure for production traits has led to loss of fitness in farm animals bred for intensive production. Serious examples include severe leg weakness in broiler chickens leading to chronic pain in birds that outgrow their strength (34), and early deaths in dairy cows exhausted by prolonged high milk production. To be fair, breeders and producers are beginning to acknowledge these problems and modify their genetic selection criteria to give less emphasis to traits linked strictly to production and more to traits linked to robustness. This has had considerable impact on the robustness of dairy cattle (78). The most extreme examples of the capacity of humans to mess around with the normal, healthy process of natural selection are displayed by what we have done to the dog. Even if the wolves that first approached human encampments had long-term foresight (which they hadn't) they could not have imagined morphing into giant Great Danes, with a high probability of dying early from bone cancers, Chihuahuas committed to repeated Caesarian sections because their heads are too large for their pelvic cavity, or Pugs and French bulldogs with noses and throats so distorted that they can hardly breath. The impact of these practices on the minds of these dogs will be considered later. Here, I would just make the point that the whole notion of breeds is unnatural. It is an entirely human construct. Adaptation to changing environments through natural selection proceeds, when necessary, through the evolution of new species. Breeding is something we have imposed on them, unasked.

Domestication, Sentience and Wellbeing

The wellbeing of a sentient animal depends on its success in adapting to its physical and social environment. In a natural environment, the animal will inevitably be faced by problems. When this happens, the animal with a sentient mind will have to think for itself. With domestication, we are able to protect them from many of the challenges of the natural world, but we restrict their independence and freedom of choice. We can also create a new set of mental stresses, often resulting not from too much challenge but too little.

The first pair of animals I consider are the dog and the pig. These may seem like an odd couple when viewed through our eyes but, in nature, the two species have much in common. Pigs and dogs (and I mean proper dogs, not man-made aberrations) are both medium-sized, strong animals with powerful teeth, capable of holding their own in combat. In consequence, both are likely to be confident, adapt well to the company of humans but can be dangerously aggressive when under stress. Both are intensely curious, keen to investigate the environment using their well-developed sense of smell. Both can be toilet trained. Both produce large litters of relatively undeveloped offspring so must exercise a great deal of maternal care, which includes aggression when they fear that their piglets or pups may be in danger. Both are omnivores: their capacity to digest food of plant and animal origin is similar to that of humans. In practice, wolves that inhabit the Northern forests must subsist largely on an animal-based diet because there is not much else to eat. Pigs can and will eat anything. Dogs and pigs also have a lot in common in their social life. In a natural environment, pigs tend to form small social groups of about three sows with perhaps five to six piglets each. Within that social group there will be a hierarchy, but sows help each other to look after the piglets and will often form communal nests. Boars are solitary. Wolves and African hunting dogs form packs made up of extended family groups with a strict hierarchical structure. Pack members work together and form close emotional bonds but recognise and are subservient to the pack leaders, the breeding pair, alpha male and female This social behaviour works very well for a species that needs to hunt as a pack in order to kill large animals like the caribou.

Pigs

Let us look first at the impact of domestication on the life and mind of the pig. The story is not all bad. I live in deepest Somerset, in the midst of a traditional mixed farm gaining its income from a sustainable mix of arable and dairy. The road between our house and the farmyard is quiet enough for dogs and pigs to roam in the street. My neighbour, Martin, breeds Large Black pigs mainly as a hobby and takes them to shows around the country. His sows range freely although constrained by farm gates. Not so the piglets, which can get anywhere. Only last Sunday, (as I write) while working in the garden I was joined by a group of eight black piglets who lolloped around with great enthusiasm, exploring this and that for about ten minutes then left for home in their own time. This, I have to concede, is atypical.

At the time of farrowing, the sow is highly motivated to make a deep bed of straw or other suitable material to create a nest in which to give birth and suckle her piglets. Forest-ranging sows make nests of branches and small twigs. This housekeeping behaviour can occupy her for many hours and given enough straw, some nests can be enormous. This is highly adaptive behaviour. New-born piglets are much more sensitive to cold than their mother and are in danger of crushing when the large, clumsy sow lies down to present her teats to her litter. A bed of deep straw keeps the piglets warm and allows them to escape when the sow slumps to the ground. Some high welfare farms reproduce these conditions very well but in the majority of intensively housed units, sows are restrained before they give birth in a metal farrowing crate that restricts their movements to standing up and lying down. A few sows may be given a token amount of straw. Piglets are encouraged to keep away from the sow and huddle under heat lamps. At three weeks of age, the piglets are prematurely weaned and transferred to barren all metal flat deck cages. For many years, the practice was to rebreed sows as soon as possible after weaning then confine them throughout pregnancy in stalls where, once again, their only options were to stand up and lie down. This was an engineers' approach to the problems of aggression and piglet mortality, but it is totally unsympathetic to the behavioural needs and emotional state of both sow and piglets. These units are also expensive to build.

In recent years, public pressure for improved welfare standards for the rearing of pigs has led to some significant improvements. Pregnancy stalls for sows are now banned in UK, the EU and many states within the USA. Farrowing crates are still permitted but, in UK, there has been a major shift towards the management of breeding sows in outdoor units with individual arks wherein they can shelter and give birth and nurse their piglets. On light, well-drained soils, these units can be as profitable as the highly engineered pig factories, especially when outdoor bred and outdoor reared pork products attract a high-welfare premium. I have heard farmers say: 'this system is all very well, but I couldn't possibly manage it because my land is far too wet'. The simple, ecologically sound answer to this is 'in which case, yours is not cut out to be a pig farm'.

In the 1980s, the UK Farm Animal Welfare Council proposed the concept of the '*Five Freedoms and Provisions*' as a simple but comprehensive framework for assessing the impact of different management systems on the physical and emotional welfare of farm animals. At the time these were defined as:

Freedom from hunger and thirst – by ready access to fresh water and a diet to maintain full health and vigour
Freedom from discomfort – by providing a suitable environment including shelter and a comfortable resting area
Freedom from pain, injury and disease – by prevention or rapid diagnosis and treatment
Freedom from fear and stress – by ensuring conditions that avoid mental suffering
Freedom to express normal behaviour – by providing sufficient space, proper facilities and the company of the animal's own kind.

These rules are close to being comprehensive and the first four *freedoms from* have stood the test of time. The fifth is a *freedom to*, and as with all such freedoms can create semantic and moral problems. What is normal? Is normal behaviour necessarily acceptable behaviour? To quote Isaac Stern '*my freedom to swing my fist stops at the point of your nose*'. On reflection, I believe the fifth freedom would be better described as *Freedom of choice* and I shall interpret it this way.

Table 10.1 uses the conventions of the five freedoms to assess threats to the wellbeing of sows in intensive and outdoor units. Whether indoors or out, breeding sows are likely to be fed once daily a ration that they can consume in a few minutes. Thereafter, outdoor sows can, if they choose, nose around for hours in the hope of finding a few discarded grains, insects or, if they are lucky, the occasional worm. Confined sows can do no more than chew the bars of their stalls. The ration of food offered to the two groups will be the same. However, the outdoor sows have had the satisfaction of foraging, the indoor sows the frustration of long hours with no oral satisfaction whatsoever. Physical

Table 10.1 Threats to the physical and emotional wellbeing of breeding sows in intensively housed and outdoor units.

Threat	Intensive, indoor	Extensive, outdoor
Hunger	High	Low
Physical discomfort	High	Low
Thermal discomfort	Low	Moderate
Disease	Low	Low
Injury	Moderate	Low
Fear and stress	Low?	Low
Freedom of choice	High	Low

discomfort will inevitably be greater for sows confined on concrete and with no opportunity to relieve discomfort through scratching or dust bathing. The risk of thermal discomfort is low for both systems. The weather can present problems for sows out of doors but, on the right land, these can be minimised through good management including the provision of shelters. When it rains, the sows can retire to their ark if they so wish. They have freedom of choice. When it is hot, they should have the freedom to wallow in mud, which is preferable to clean water since it evaporates slowly and so provides a lasting cooling mechanism for pigs who cannot sweat. White pigs can suffer from sunburn, but breeders now produce hybrid sows with pigmented skin better suited to being out in the sun. These strains are usually derived from sows of the coloured Saddleback, or Hampshire breed, which are proven good mothers and therefore particularly well suited to environments where the responsibility for successful piglet rearing has been left largely to their care. Cold stress is seldom a problem for sows, especially if they have access to a well-insulated bed in their arks. Cold, dry winter days do not appear to restrict the time they spend foraging out of doors. The risk of injury is set at moderate for indoor and low for outdoor systems. Sows condemned to live on concrete are more prone to abrasive injuries and musculoskeletal problems arising from their restricted movement. Sows can be aggressive and do a great deal of damage to one another especially when in close confinement. Outdoors, at lower density, they will usually sort out problems by social distancing. The occasional rogue sow can be removed.

The risk of fear and stress is set at low in both systems, although in confined systems, this is qualified by a question mark. Aggression in growing pigs can be a problem, especially when they are densely stocked in barren environments. When sows or growing pigs are housed together, it is now customary, and in some places, compulsory to provide 'toys'; manipulatable objects like chains or tires for the pigs to play with and thereby reduce the risk of fighting. EU regulations state that *'pigs must have permanent access to a sufficient quantity of material to enable proper investigation and manipulation activities such as straw, hay, wood, sawdust, mushroom compost, peat, or mixture of such that does not compromise the health of the animals.'* This is just as it should be, but I suspect that neither the means nor the objective are regularly achieved in practice. Pigs are extremely curious, highly motivated to explore and forage, but they do this, quite reasonably, in the hope of a reward. The attraction of a dangling chain soon wears off. Sows confined in pregnancy stalls are freed from aggression and the fear of aggression but may develop a chronic state of depression due to complete denial of the freedom of choice to make a positive contribution to their own wellbeing. All pigs are powerfully motivated to explore their environment with the most lasting stimulus being the anticipation of a food reward, however small. When sows are housed in groups on deep, clean straw, their daily ration of pelleted food can be scattered randomly over the entire area. This keeps them occupied for hours and given time, they will nose out every last pellet.

Dogs

After millennia of domestication, dogs retain many of the emotional and behavioural traits of their wolf ancestors. They gain security and pleasure from life in stable family groups, and each is powerfully motivated to make its own active contribution to the overall wellbeing of the pack. They strengthen these links through close emotional attachments within the family and show the greatest of respect to the alpha breeding pair at the top of the hierarchy. All these traits adapt well to life in human families so long as the human family follows some of the basic, simple codes of the wolf pack. The dog must identify and respect a clear leader. As in wolf packs, this can be a breeding pair. The dog should learn to live, and enjoy life, with other dogs, both within and outwith the family. The process of socialising with other dogs should begin when the puppy is about 8 weeks of age, as soon as possible after routine vaccination. One feature of canine socialisation that may appear strange to human eyes is that all dogs, from dachshunds to Great Danes instantly recognise one another as dogs. This is a powerful illustration of the fact that each species selects and develops its own set of special senses to establish social contacts. For dogs, smells are likely to be the most important. Visual information will obviously include size, but this will be seen as irrelevant unless linked to threat, conveyed by body language in the form of signals such as bared teeth, posture and laid-back ears.

The powerful need of dogs to make a positive contribution to family life needs to find expression in some form of activity. We and they can put this to good use by employing dogs in a variety of useful roles that range from the highly active sheepdogs and sniffer dogs to the patient guide dog. The family pet that carries a stick when out on walks or chases balls for hours may not be achieving much but it is fulfilling the same motivational need to contribute to family life and derives great satisfaction from so doing. To my knowledge, dogs are the only species that will work hard not in the expectation of an immediate tangible reward like food but simply to express and reinforce ties of mutual love within the family.

In the right circumstances, dogs make wonderful companions. We love them and they love us. Love is a word to be used sparingly and with great care. I cannot apply it with any confidence to the behaviour of any other animal with respect to their human acquaintances. Indeed, why should they love us? Cats enjoy human company, which includes the food and comforts we provide. They also display a great deal of what we would call affection, which typically involves climbing into our laps to ensure that we stroke them. Well-adjusted pigs will solicit a good scratch. Horses, to be discussed later, have a hierarchal herd structure, somewhat similar to dogs, and show respect to a recognised leader, who may be human. Dogs may be unique in their motivation to contribute so much without expectation of a tangible reward. However, I would not go so far as to call this true altruism. The motivation for these actions is to contribute to the wellbeing of the extended family according to the JBS Haldane principle that he would offer to lay down his life for two of his immediate siblings or eight of his second cousins. Dogs living with humans do, at the deepest level, consider themselves to be one of the family.

The strong hierarchal and emotional link between dogs and their human families has its downside. Abnormal behaviour, indicative of a disturbed emotional state, is a big problem for pet dogs. We see evidence for this in signs of overexcitability, such as excessive barking and jumping up on people, aggression to humans and other dogs, and destructive behaviour, e.g. chewing up the furniture. What we may fail to see (because we are not there) are the signs of separation anxiety, which is the most common expression of mental distress in dogs. Most publications indicate that about 40% of dogs attending veterinary surgeries are brought in with behavioural problems. This underestimates the true incidence because a large number of these dogs are not referred to vets. Behavioural problems account for about 10% of all cases for euthanasia. This is not the place to explore in detail the origins and treatment of behavioural problems in dogs. It is, however, the place to point out that all can be linked to the fact that the balance of their mind has been disturbed through exposure to a series of social signals alien to their genetically built-in model for mental formulation. These include lack of a clearly recognised pack leader, confusing or inconsistent signals from the assumed pack leader (too many words, not enough expression, erratic body language), failure to establish socially acceptable relationships within the pack, with other animals (dogs and humans) outside the pack and, especially, prolonged isolation from other pack members and anxiety as to when (or indeed, if) those bonds will be re-established. The message is simple. Loving dogs is not enough. We have a duty to offer them the sort of life that they can understand; in short, a proper dog's life.

Cats

As I wrote earlier, cats have permitted themselves to become domesticated on their own terms. When they have freedom of choice to roam, hunt, and select the most comfortable spot in the house, they have a great life. A number of cats are referred to vets with behavioural problems. Many of these, like spraying urine or aggression to other cats are not really problems at all, in cat terms, since they are expressions of normal social behaviour. Kipling was spot on when he wrote 'I am the cat that walks by itself'. Cats, like tigers, are solitary animals. While they can form close bonds of friendship, other cats are initially viewed as competitors. Similarly, spraying is a demonstration of scent marking by a cat seeking to establish its own territory. Some cats will exhibit patterns of stereotypic (compulsive) behaviour like chewing wool or fabric. These are more likely to occur in cats that are kept permanently indoors. Abnormal behaviour patterns linked to emotional distress are much less common in cats than in dogs, because they are so much less emotionally dependent on us, and most have so much more freedom.

I make one big exception to the general rule that domesticated animals should be given as much freedom as possible to express their natural behaviour. I am convinced that the quality of life of the domesticated cat is improved by neutering them to eliminate their innate patterns of sexual behaviour. Female dogs only come into season for a few days, once or twice a year. They are usually kept close to home, and it is relatively easy to manage their sex lives with little stress to either party. Oestrus in female cats is likely to be prolonged until they conceive as a result of a successful mating. Moreover, they can return to oestrus while still feeding their litter from a previous mating. Female cats, overworked by the stresses of motherhood, become short tempered and exhausted. Entire male cats, free to roam at night, get into fights and can suffer serious injury. I believe that it is an act of kindness to detach them from this hard-wired passion.

Dairy Cows

The sentient minds and social structure of cattle at pasture and on the range were discussed in Chapter 8, along with other grazing animals. The individual dairy cow is a special case, to be considered with our other close neighbours because she, of all the domesticated food animals, has lived closest to us and contributed most to human well-being over the last 10 000 years. The word cattle has the same origin as chattel and capital, descriptions of personal property and wealth. The expression 'cash cow' is used to describe an inexhaustible money-making machine. Throughout most of the history of agriculture, and in much of the world today, the role of the family cow has been to provide us with a wide range of goods; milk, work, fertiliser, fuel, clothing and the occasional fatted calf, while sustained by fibrous feeds that the family could not digest for themselves, much of it from land that the family did not own.

The modern dairy cow, typified by the Holstein breed, is bred, fed and managed to produce as much milk as possible while housed intensively in large barns and milked using highly mechanised dairy units. The sole purpose of these highly specialised units is to produce as much milk as possible as cheaply as possible. Meat production has become a minor consideration, with calves destined for beef or veal sent, more often than not, to other specialist rearing units. Surplus male dairy calves may be killed shortly after birth because they have no value. Other traditional roles for the family cow have disappeared altogether. The modern Holstein is most unlikely to be harnessed to a plough!

When measured strictly in terms (short terms) of economics, these large industrial units are an undoubted success. When measured in terms of sympathetic and sustainable husbandry of the land and the animals, they are found to be wanting. The needs that drive the mind of the modern, highly bred, intensively fed cow are much the same as for any sentient mammal: food and water, comfort, security and a stable social life consistent with the genetic imperative for sex. Fundamental to all these specific needs is freedom of choice: to take action to avert discomfort or threat and promote a positive sense of wellbeing. As we know too well, the impact of food on our state of mind is not just a matter of acquiring sufficient nutrients. So too with cows. The acts of eating and, in their case, ruminating, bring their own satisfaction. Grazing animals in the wild state have adapted to seasonal changes in food availability: lots of good grass in the summer or rainy season, much less food of far poorer quality in the winter or dry season. It is entirely natural for grazing animals to lose weight during the lean months, but provided some grazing is available, however poor the quality, they get the satisfaction of freedom to forage for what they can.

The high yielding cow in a modern intensive unit is presented with a quite different problem (79). Cows are not motivated to eat by a desire to reward the farmer with as much milk as possible, but by the desire to attain a feeling of comfortable satiety. High yielding cows experience a sense of chronic hunger driven by the demand for nutrients to sustain the enormous quantity of milk they produce. At the same time, their capacity to take in food, especially fibrous food essential for healthy digestion, is constrained by the rate at which this food can be fermented in the rumen. Energy-rich rations designed to increase milk yields by increasing the proportion of rapidly digestible starch to slowly digestible fibre can lead to indigestion and metabolic problems such as acidosis, which make the cow feel ill. In consequence, many high yielding cows can simultaneously feel hungry, full up and liverish. This is not a good feeling.

Cows' need for comfort is greatly influenced by their size and shape. The modern Holstein weighs over 700 kg and has prominent joints, especially knees and hocks. For comfort, they need to lie down on pasture or a deformable bed of straw or sand. Concrete does not feel good. Cows are motivated to lie down to rest for about eleven hours per day. There comes a point where the need to lie down overrides the need to eat. In many intensive units, high yielding dairy cows are milked three times daily, having queued to enter the milking parlour. They are also compelled to eat for at least eight hours to meet their nutrient demands. With so much to do, the time to lie at rest will be much less that they would wish.

Cows, like all sentient animals, are motivated by curiosity and caution. Curiosity is a powerful motivator in early life as calves seek to gather useful information. In later life, in a stable environment, caution becomes the wiser approach to ensuring a sense of security. Most cows in stable groups establish a stable hierarchy, through the exchange of social signals that usually avoid physical conflict. In houses where each cow has access to an individual cubicle, it is normal for each to use the same cubicle every time. Overworked by the demands of lactation, they opt for the quiet life. However, they do retain their curiosity. If you wish to be entirely surrounded by curious cows, lie down in a field and the rest will follow. Horizontal, we present no threat and become interesting (Figure 10.1).

Figure 10.1 The author relaxing among friends

Whether on the family farm or in large intensive units, the dairy cow is a valuable individual and will be given individual attention. Despite this, dairy cows are at high risk of three major health problems, infertility, mastitis and lameness. These conditions are known as production diseases, a phrase that concedes that they are largely our fault. Farmers do their best to minimise the incidence of these conditions since it is in their own best interests. There are, however, some practices that we inflict on cows entirely for our benefit, in full knowledge that they conflict with how they would naturally perform to promote a sense of wellbeing. The top three, in ascending order of importance, are:

- Tethering cows throughout the time they are housed
- Keeping cows permanently housed, without access to pasture
- Removing calves from their mothers shortly after birth.

In many small rural communities, it has been traditional to keep dairy cows outdoors all summer on lush pastures, like Alpine meadows, then bring them in for the winter and tether them in tie-barns where they will be fed, watered and milked until turn-out in the spring. This practice has given rise to concern mainly on the grounds that it denies freedom of movement and opportunities for a social life. I know of no evidence that cows display signs of distress associated with prolonged tethering, although passing the winter group housed in a barn with deep clean straw and access to an outside yard would undoubtedly be better. Some free-stall houses with insufficient, poorly bedded cubicles and filthy passageways can be worse than tie-stalls. In any event, tie-stalls are incompatible with modern milking systems and will, I predict, gradually fade away.

The trend in commercial dairy production, worldwide, is towards very large, industrialised units of 1000 cows or more very high-yielding cows. In order to sustain these high yields, the cows are housed throughout lactation and given continuous access to a ration that ensures they take in far more nutrients than they could possibly derive from grazing at pasture. Confinement also keeps them close enough to the milking parlour to permit thrice-daily milking or, increasingly, the use of a robot milking machine that they enter of their own free will. Cows are not highly motivated to be milked *per se* so need a food stimulus to attract them in. Robot milking machines are only practicable when cows are permanently housed because they are reluctant to come in off pasture. While all cows are confined throughout lactation, some large systems allow cows access to pasture for a few weeks during the dry period when they have completed their lactation and await their next calf. In the UK, we are accustomed to seeing dairy cows outdoors at grass during the summer, so Brits assume this to be the natural state. However, this is becoming the exception through most of the developed world where the majority of lactating dairy cows are kept off pasture throughout their working life.

We humans differ widely in our opinion as to what we feel is right for cows. The real question, of course, is how do they feel about it themselves? Pasture is great. Here, they can do much as they please: take in food, dump urine and faeces, exercise, rest, enjoy fresh air and space, socialise and satisfy their curiosity. Pasture has good points for the farmers too. Cows can forage for themselves, and their manure doesn't have to be scraped. However, too many cows excreting and walking about in mud wrecks the pasture as a source of food. Once the first flush of high-quality spring grass is over, many farmers turn their cows out onto sacrifice pastures that provide little grass but all other amenities. The only real benefit to cow and farmer from high-quality pasture is easy access to high quality, highly palatable food. This is a big plus but there are other ways of achieving the good life. The most cow-friendly farm I have ever seen was in the forested foothills of the Pyrenees in northern Spain. Cows could choose to roam in comfort and security among the trees, or rest in well-bedded kennels. There was little to eat in the forest. All food, including freshly cut grass in season, was provided at a feeding station close to the milking parlour. This system came as close as possible to meeting all their day-to-day needs, but it was exceptional. However, even in the largest and most intensive systems, it should be possible to ensure that cows have freedom of choice to go outdoors, when they wish, where there is space, cool fresh air and a comfortable place to rest. I have to conclude that this does not necessarily equate to a field of green grass. Nevertheless, I shall always be moved by watching the joyful antics of cows when, after a long winter, they are first turned out in the spring.

The third and most serious of our insults to the emotional state and natural behaviour of dairy cows is the policy of removing their calves shortly after birth, partly for ease of management but mainly to maximise income from sale of milk. Whether in the wild, or out-of-doors on the farm, the natural behaviour of the dairy cow at parturition is to separate from the herd and give birth in what she thinks will be a safe spot, for example, close to a hedge. Having licked the calf into shape and given it a first meal, she leaves it and returns to the herd; instructing it, in effect, to lie still and unnoticed until she returns to give it another drink. This behaviour is hard wired and has survival value. After a few days, when the calf has become active and can move as well as its mother, it will join the herd, spending much of the time with other calves, because they are more interesting, visiting its mother perhaps 4–6 times daily for a feed and usually resting with her at night. It is natural for cows and their calves to spend a long time apart, but both show signs of distress if not together at mealtimes. A few farmers separate cow and calf but allow the calf to join its mother twice daily to take a modest feed before the rest of the milk goes into the machine. This system appears to be acceptable to both mother and calf. Many domesticated water buffalo, e.g. in India, will not permit themselves to be milked unless their calf is present.

While I believe that the twice-daily access system is a reasonable approach to good husbandry, it is likely to remain a minority pursuit. What then is the least-worst approach to early weaning? In this context, the French word *sevrage* is more accurate. At present, the most common practice is to separate the calf within 24 hours of birth. On some traditional dairy farms, calves will be left with their mothers for 2–3 weeks to ensure they get a good start in life. Weaning after three weeks undoubtedly causes more distress to cow and calf than weaning shortly after birth. Both will continue to bawl for each other for hours or days. Early weaning is an unpleasant business but, in the words of the murderous Macbeth '*when 'tis done, it were well it were done quickly*'.

Horses and Donkeys

Horses, donkeys and zebras are the three living species within the genus Equidae. Horses (mostly) evolved on the steppes of central Asia, donkeys and zebras on the plains of Africa. Horses have adapted reasonably successfully to domestication. Zebras have refused. Donkeys have been too stoical for their own good. Horses were first 'broken in' about 5000 years ago and put to work pulling chariots and carts. We developed our riding skills a little later. I find it odd that only three species of Equidae have survived in the natural state compared with over 200 species of ruminant. Both genera are herbivorous prey species that live in herds and rely on one another for security from predation. Here, we need to enquire what was special about the environment that favoured the Equidae over the ruminants and the consequences of this form of evolution on their bodies, minds and adaptation to domestication.

The most fundamental difference between the Equidae and ruminants is in the design of the digestive tract. Both genera have evolved on diets consisting mostly of grasses and other fibrous foods that can only be digested by microorganisms inhabiting a large fermentation vat at one end or other of the digestive tract. In ruminants this is at the front end, the rumen; in horses it is at the back, the caecum and colon. The feeding pattern of ruminants is to consume a large quantity of food as quickly as possible then retire to a safe spot, perhaps among the trees, to regurgitate and ruminate on it at leisure. The small stomach of the horse means that it cannot consume large meals. In the wild state, it needs to graze out on the open plain for up to 13 hours per day. This exposes it to greater risk from predators. To compensate, it has, through natural selection, developed the capacity to run and run. The early horse of the steppes, the progenitor of the Arab and Thoroughbred, had the build, the heart and lungs of the athlete. Moreover, the horse that ran the fastest was the least likely to get caught.

The social structure of wild or feral horses is hierarchal. All 'wild' herds today are, strictly speaking, feral, having escaped domestication. A typical herd may consist of eight mares, their foals, and perhaps two stallions, who tend to stay on the periphery, except when a mare is in oestrus and receptive to mating. The leader of the herd is usually an alpha female, who makes the decisions as to where they should graze and when they should move on. While the structure is hierarchal, it is usually stable and maintained with minimal aggression. The establishment of social bonds is achieved through a small number of visual signals, position of ears, angle of head, proximity and angle of approach. When approached by a younger, less well-established horse or foal, a dominant mare will, at first, display signs of dominance and mild threat. The new horse, lower in the hierarchy, will respond with signs of submission such as lowered head and chewing motions. The dominant mare acknowledges these signs and offers positive reinforcement to the newcomer by withdrawing the threat signals to the point where it can relax in the confident belief that it has been accepted as a member of the herd – so long as it remembers its place.

The behaviour and social structure of donkeys are similar but not identical to those of the horse. When there is sufficient grazing, they form groups of about 6–8 mares with one or two stallions (Jacks). The lead stallion acts as herd leader. Their normal response to the threat from predators is to run away. Since they cannot run as fast as horses they may be forced to turn and fight. In times of food shortages, individual mares and foals may separate from the herd and seek their own salvation. Despite this, donkeys form close emotional attachments within the herd. Domesticated donkeys display chronic signs of distress if kept in isolation. Donkeys that have lived with a particularly close friend show extreme distress if separated by death or otherwise and may not be consoled by the arrival of a new companion. Some may just pine and die.

The messages to be taken from this brief examination of their natural herd behaviour is that horses and donkeys are motivated to eat fibrous foods for long periods, and they respond better to positively reinforcing signals than to punishing displays of dominance. They experience distress when isolated from the herd, or individuals within the herd. All these expressions of feeling are critical to their ability to adapt to the lives we impose on them through domestication. Horses kept by us for recreation, sport, or just as pets, can suffer from a range of behavioural disorders directly attributable to the abnormal lifestyles that we have imposed on them. These have been described as 'stable vices': a cruel misnomer, as they are almost always our fault. The phrase 'stable vice', like 'breaking in', should be confined to history.

Two expressions of abnormal stereotypical behaviour, weaving and box-walking, are caused by the stresses of isolation and confinement. The weaver typically stands at the stable door and compulsively weaves its head from side to side. The box-walker compulsively walks round and round its stable, pausing at the door as if in the expectation that this time it might just be open. The engineers' solution to these problems has been to install anti-weaving bars on the door, to make weaving impossible or to close the door altogether. There is dispute as to whether the stereotypical actions of weaving and box-walking (akin to head banging in deprived children) are expressions of chronic distress, or a form of adaptation that serves to reduce distress. Either way, they are evidence of deprivation. If they are a means of relieving stress, then the installation of anti-weaving bars makes matters worse.

The other common expression of stereotypical behaviour in horses is crib-biting combined with wind sucking. Typically, the horse grasps the top of the stable door between its teeth, and gulps in air, not to the lungs but into the oesophagus in a sort of reverse belching action. This behaviour was assumed for many years to be an expression of frustration or, at least, boredom. Recent studies show that it is primarily due to improper feeding (47). I wrote earlier that the four essential criteria of a good diet are that it should provide the right nutrients, promote healthy digestion, promote oral satisfaction and do no harm. Many rations fed to stabled horses fail on three counts out of four. A horse that gets most of its ration in the form of one or two high-starch concentrate feeds per day, supplemented with a handful of hay will clearly lack the oral satisfaction to be gained by grazing or foraging quietly for several hours. Moreover, such feeds, especially when fed to newly weaned foals can lead to painful ulcers through excess production of acid in the stomach. When we suffer from acid indigestion, we take tablets containing buffers, especially sodium bicarbonate, which is a natural constituent of saliva. Horses, unlike ruminants, do not salivate all the time, only when they are actually masticating and swallowing food. What crib-biters are doing is replicating the actions of mastication and swallowing, thus stimulating the flow of saliva and easing the discomfort of acid indigestion. This is an adaptive behaviour so any attempt to prevent wind sucking by fitting a tight collar round the horse's neck will make things worse. Feeding a ration that meets all four criteria of a proper diet will greatly reduce the risk of the horse becoming a crib-biter. However, this, like all stereotypic behaviour is hard to eliminate once it has become established.

Our greatest insult to the horses we love but fail to understand has been the traditional process of 'breaking in' to the saddle and rider or the shafts of the cart. This was traditionally based on the principle of establishing dominance through fear of punishment. At its worst, it involved presenting the horse with a series of aversive and frightening challenges, which it fought against to the point at which it finally submitted through sheer exhaustion. Over the years 'breaking in' methods have become less severe, while still essentially based on the principle of dominance through punishment of 'bad' behaviour. A new and completely different approach to horse training has been pioneered by Monty Roberts. He calls this 'joining up'. This is best explained through watching it on YouTube. A novice horse enters the ring. Initially he (Monty Roberts) presents the mild threat of a whip that doesn't touch the horse, or just a sudden movement, that the horse can easily avoid by running around the ring, not in panic, but as an unflustered form of avoidance behaviour. The horse will soon start to display the same submissive gestures that it would display in the herd, like lowered head and chewing movements. Monty then responds in horse-like fashion, positively reinforcing these signs of submission, by replacing the signs of threat with those of acceptance. Within minutes, he stands in the middle of the ring, the new horse approaches him, 'joins up' and he responds with an affectionate stroke of the head. There are two particularly interesting features of this demonstration. The first obvious one is the speed at which the horse develops a stable, friendly relationship with the new human alpha member of the herd. I find the second to be even more impressive. At each stage of the process, Monty Roberts says 'I will now do this and then the horse will do that', and every time he is correct. What he is demonstrating is that the visual signals necessary to establish stable social relationships, whether within a group of horses, or between horses and humans, are simple and few. The simple message is quickly received and understood. The horse does not need to expend too much thought to conclude that while both parties follow the rules, it knows and will remember how to behave in such a way as to enrich its quality of life with an added sense of security.

The word 'remember' sets me off at a tangent to the training of elephants. Traditionally, this involved the most severe form of breaking-in through punishment. Elephants were caged or chained to a tree and whipped as they struggled to escape until they were exhausted. While working elephants may subsequently be treated quite well, not least because they are highly valuable, we may assume that they have been motivated primarily by the fear of future punishment. However, as the saying goes, elephants never forget, and occasionally run riot. The social structure of elephant herds is similar to that of horses, a stable hierarchy under the benign leadership of a dominant female. There has, in recent years, been a movement to train elephants using a version of the Monty Roberts 'join up' approach. These are early days, but I am advised by an elephant whisperer that, to his knowledge, no animal thus trained to social life with humans has, so far, gone rogue.

Chickens

Today, the vast majority of hens kept for commercial purposes in the developed world are housed either in cages or very large commercial units that severely restrict their natural behaviour. The behaviour and welfare of hens in intensive units has been a matter of great concern for many years. Thanks, in some degree to the diligent studies of compassionate scientists but, in greater degree, to the pressure of public concern, there has been a progression away from the battery cage towards more free-ranging systems, still operating on a very large scale. This is mostly a good news story but outside my current remit.

In the most intensive units for production of eggs and poultry meat, the birds have little, if any, opportunities to exercise their minds. In order to understand the mind of the chicken, it is best to study them in small groups within an enriched environment that permits them to express their individuality, such as an African village, a traditional British farmyard, or a suburban garden. The most valuable birds are the laying hens: the egg producers. Chickens are one of the very few birds that we have been able to domesticate for the simple reason that they find it hard to escape. The modern hen is a descendent of the jungle fowl of south-east Asia. These birds have a very limited ability to fly. During the day, they scavenge the jungle floor for a diet that will contain a high proportion of insects and other invertebrates, protected by the forest cover from predation by raptor birds. At dusk, they will make the short flight up to roost on a low-lying branch to get protection from predators on the ground. They lay large eggs rich in protein and fat to feed a developing embryo outside the body of its mother to the point of hatching. In the wild, where their eggs will be fertilised by a cockerel, they may lay six eggs, then brood them to the point of hatching. When their eggs are removed, they will continue to lay eggs on a daily basis for many months. All these traits made them ripe for domestication.

In a traditional African village, the family that owns a few hens can provide their children with a highly nutritious egg perhaps twice a week, kill and eat a bird on very special occasions and, ideally, get some cash income from egg sales. For many years, the opportunity for mothers of families to improve their lives from small-scale egg production was limited by infection with fowl pest, the Newcastle Disease virus, that would destroy the flock during the rainy season. A number of charities are now providing funds to enable the vaccination of village chickens against fowl pest, thereby allowing families to maintain healthy flocks of about 15–20 birds on a year-round basis. When I enquired 'but where do the ladies get the money from to provide feed for their birds' I was met with a laugh. 'Oh, they don't feed them'. So long as numbers don't get too high the birds can scavenge for themselves, their diet including large numbers of insects that would otherwise be a burden to humans and other farmed animals. These birds have become highly valued members of the family while retaining a pattern of behaviour essentially identical to their wild ancestors.

Hens are one of the most fascinating animals to observe because they are always doing something. They spend long hours investigating almost everything, mainly in the expectation that it might be eatable. Having no hands, they do this with their beaks. Groups of hens establish and defend a strict hierarchy, the 'pecking order'. Aggressive pecking between hens is usually directed to the head and neck. Cockerels will fight to the death (and men bet on the outcome). One of the original justifications for confining hens in cages was to prevent the most severe consequences of feather pecking. Problems can arise in some large commercial flocks where hundreds of birds can be plucked bare over large regions of their body. The motivation underlying feather pecking is not fully understood. Large-scale feather pecking does not appear to be associated with aggression. The beak is the tool that hens use to express their curiosity. In close proximity, and in the absence of other interesting things to investigate, hens may peck at and pull feathers from one another at any point of the body without overt signs of aggression from the pecker or submission from the peckee. When small, stable groups of hens are free to range in an enriched environment with plenty of places to explore in the expectation of a food reward, feather pecking is rarely a problem.

In the UK, it has become commonplace for people who wish to keep a few hens, partly for their eggs and partly because they are fun to have around, to get 'end-of-lay' birds from commercial units, especially those keeping their layers in cages. When these birds arrive, they are physically weak and totally naïve, having had no experience of a natural life. However, they adapt quickly and, within weeks, appear to show nearly all the skills necessary for independent existence. However, they appear to display less caution than jungle fowl, or birds raised from the moment of hatching in a risky environment. They do seem to be an easy prey for foxes. This suggests that they are lacking the benefits of a proper education. I have already described how chickens build on their birthright and develop their minds largely from observation of other birds. They spend more time observing dominant birds in the hierarchy on the basis that to adopt their ways is the route to success. Moreover, their welfare can be enriched by formal education from their mother hens who communicate survival skills to their offspring on the basis of knowledge acquired in their own lifetime. Social sentient animals with few natural defences need the benefits of a good education.

Opportunist Neighbours: Rats and Urban Foxes

So far in this chapter, I have been considering what goes through the minds of the very small group of animal species that have permitted themselves to experience domestication, largely on our terms. Before closing however, I must consider some of the species that have been compelled to adapt, or chosen to adapt to civilisation, seen through their eyes as environments dominated by human development and human behaviour.

The rat is a species that is identified as vermin and persecuted accordingly yet thrives where the density of the human population is at its highest. It provides the most striking evidence in support of the premise that the greater the challenges and complexity of the environment, the greater the range of skills the animal must develop in order to cope. The range of innate and learned skills demonstrated by rats in laboratory experiments is legion but here is not the place to review them. Rats share with humans the capacity to first dominate, and ultimately wreck the environment unless kept under strict control. Like humans, they have colonised almost the entire planet, travelling by ship when necessary. When introduced to a previously rat-free habitat, they can cause devastation, for example, by stealing the eggs of ground-nesting birds like the puffin.

The great majority of brown rats today are entirely urbanised, raising their families in the security of the sewers, establishing strong social bonds within families and displaying aggressive resistance to outsiders perceived to be a threat. They come out, mostly at night to scavenge in areas where they know food has been discarded and, when necessary, can chew their way through most containers. They are prepared to sample anything, which puts them at high risk of poisoning and infections. Much of the resistance of rats to infection is likely to be the result of harsh genetic selection: the rats of today are the offspring of the few resistant survivors of early epidemics. However, there is clear evidence that rats can learn to avoid poisons even before they have tasted them for the first time (12). For many years, warfarin was the rat poison of choice. This is an anticoagulant drug that has a cumulative effect. Rats apparently showed no aversion to foods containing warfarin but as the poison accumulated, died from internal bleeding. The UK Pesticide Safety Directorate has described this process as 'markedly inhumane' (74). During the warfarin era, a group of super rats emerged that not only rejected foods dosed with warfarin but did so when the food was presented for the first time. They had discovered in advance that this food was dangerous. Rats, like dogs can not only smell sickness but identify its cause. We assume they got this information by associating the smell of other sick rats in the early stages of warfarin poisoning with the smell of the treated food itself and conveyed this information to other members of their family. We may assume that rats, like mother hens, acquire survival skills in their own lifetime and pass this new knowledge on to their offspring, thereby enriching the mental capacity for survival within the population.

Some years ago, Manuel Berdoy took a population of rats bred for hundreds of generations in the controlled and almost entirely barren environment of the laboratory and introduced them to a natural, but enclosed forest environment in Wytham woods outside Oxford. Within days, these rats successfully displayed all the basic skills necessary for survival: foraging, comfort, shelter, security (6). Like battery hens rescued from battery cages and introduced to the challenges of the world, they demonstrated that they had retained the genetic, hard-wired instruction manual inherited from their distant ancestors. However, like the chickens, they will have been naïve, lacking the benefits of an education from better informed adults and so at a disadvantage relative to rats habituated to the wild.

Other species that have benefited from urbanisation include the urban fox in the UK and the racoon in North America. In both cases, these animals have profited from the fact that town and city dwellers are more inclined to be sympathetic to these pretty animals than subsistence farmers and, more critically, throw away much more food. In consequence, urbanisation has been beneficial to these species, measured simply in terms of population numbers. However, it is impossible to escape the conclusion that these animals have demonstrated their freedom of choice and moved into the cities to promote their own wellbeing.

Coda

This chapter has sought to interpret the effects of extreme domestication as experienced by the animals themselves. Our treatment of them is highly influenced by their utility to us. Pets are loved as individuals. Farmyard animals are treated in a more utilitarian fashion, but in recognition (we hope) that they are not simply commodities but sentient species. In the case of both groups, there can be a form of social contract wherein each party gives up certain freedoms in return for certain benefits. The lives of most farmyard animals will be short and frequently uncomfortable, but they do have the social benefits of living among their own species. It is an inconvenient truth that the animals we love the most, dogs and horses, are those most likely to exhibit signs of abnormal behaviour associated with emotional distress. It is unnecessary to invoke the words of Philip Larkin to make the point that love, and understanding are not necessarily synonymous.

Part 3

Why it matters: Nature's Social Union

'I'm truly sorry Man's dominion, has broken Nature's Social Union
and justifies that ill opinion, which makes thee startle,
at me, thy poor earth-bound companion, an' fellow mortal.

Robert Burns 'To a Mouse, on turning up her nest with a plough.

Our Duty of Care

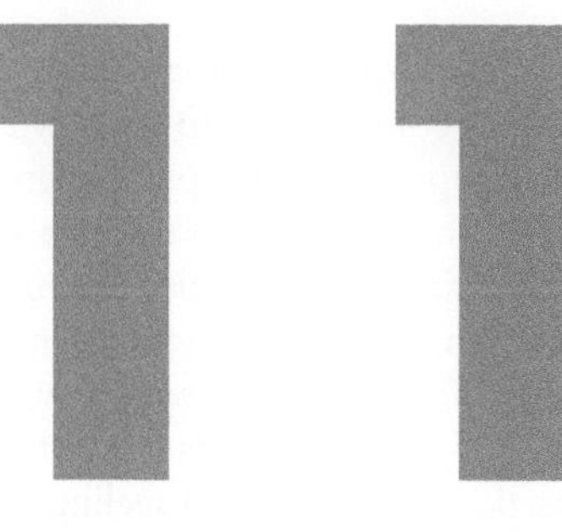

If we are to meet our duty of care to other sentient animals, we must try to get inside their heads. In pursuit of my central aim, 'Understanding sentient minds' I have done my best to look at life as seen through their eyes (and other special senses) and how these impressions and images will direct behaviour necessary to cope with challenges to their welfare. In this study, I have sought to distinguish between instinctive responses built into their birthright and those that reveal properties of a sentient mind that can not only experience emotions of suffering and pleasure but also, to a greater or lesser extent, *understand* them.

I now address the second clause in my title 'Why it matters'. We must address the evidence that our fellow mortals are, in the elegant words of Judith Benz-Schwarzburg and Andrew Knight, our '*cognitive relations yet moral strangers*' (5). As in my first animal welfare book, 'A Cool Eye towards Eden', I take as my inspiration the words of Robert Burns. *'I'm truly sorry Man's dominion, has broken Nature's Social Union, and justifies that ill opinion, which makes thee startle, at me, thy poor earth-bound companion, an' fellow mortal.'* He expresses true compassion at destroying her nest: '*that wee bit heap o' leaves and stibble, has cost thee monie a weary nibble*' but is equally aware that he cannot stop the plough for he too *'maun live'*. I repeat what I wrote then. '*We have dominion over the animals whether we like it or not*'. If we are to do right by them,

Animal Welfare: Understanding Sentient Minds and Why it Matters, First Edition. John Webster.

we need to examine our own attitudes and especially our behaviour through their eyes. It is of no real consequence to them what we think, it is what we do that matters.

The first step towards proper recognition of animals as fellow mortals is to acknowledge that we share the properties of a sentient mind. Whatever we might think, we are all driven primarily by complex feelings. To meet our duty of care, we must do more than attend to their obvious physical and behavioural needs, we must address our thoughts and feelings to the way they think and feel. For those animals not directly in our care but for whom we bear responsibility as stewards of the living environment, we must take their thoughts and feelings into account in any strategy for environmental management and population control. Specific strategies may range from actions designed to conserve an endangered species to a policy of population control for a species that is endangering its own welfare by exceeding the provisions of its habitat. In all cases, we should aim for a just balance between potential harms and benefits for them and us.

The pursuit of justice for animals requires us to adapt the principles of Hobbes social union (or social contract) for human society to incorporate not just sentient animals but all life. Humans in the social union include those directly involved as, for example, farmers or pet owners, and those indirectly involved as consumers of animal produce or competitors for habitat: in short, everybody. Animals in the social union include those directly under our care and those for whom we have no legal duty of care but who are, nevertheless, within our dominion because we have so much influence over where and how they live. The third involved party includes everything else that contributes to the quality and sustainability of the living environment, the plants, the water, the soil, the air.

A modern, democratic interpretation of Hobbes' concept of the social contract would state that the quality of life for all individuals depends on our commitment to life in an ordered, fair and functional society where each individual has a responsibility to contribute to the general welfare and accept some restrictions on personal freedom. This is often interpreted as respect for equal rights. I steer clear of the debate on animal rights for the good reason that it is a debate carried out entirely among humans; our fellow mortals are outside the room. In any ethical analysis of our relations with sentient animals and the living environment, we must make a distinction between Moral Agents, the humans who make the decisions, and Moral Patients, the animals and the living environment whose wellbeing and autonomy are affected by our decisions. We have a responsibility to them, but they have no responsibility to us. In this sense, they are like newborn babies.

Sentience Revisited

At this stage, I think it would be helpful to recapitulate my approach to the understanding of animal sentience according to the five circles (skandhas) of Buddhist philosophy, which describes all animals as sentient but some more sentient than others (*see* Figure 2.1). I argue that the five skandhas create a practical, scientifically valid set of categories into which we can put different, more or less complex properties of mind revealed from observation of animal behaviour and motivation to behaviour. I must concede, however, that these divisions are not clear-cut and, in many cases, we have too little understanding of animal minds to conclude with any certainty which properties of sentience can be attributed to which species.

According to the Buddhist skandhas, the outer circle of sentience is *Matter,* a property of all lifeforms, plants and animals. It embraces the ability to respond to signals such as light, temperature and the physical properties of the immediate environment of air, land or water. The principle of respect for life requires that matter (as defined by the skandhas) matters. If we take this for granted, the pragmatic next question becomes; '*How much does it matter*'? At this point, ethical purists may charge me with the sin of moral relativism, but so be it. It matters because our privileged position as stewards commits us to do the best we can, to ensure the survival, complexity and beauty of the living world. In this context, the best advice must be, wherever possible, leave it alone. This does not mean that we are obliged to strive to keep every living thing alive for as long as possible. However well meaning that, would be catastrophic.

The second level is *Sensation.* Animals who demonstrate the property of sensation *but no more* recognise and respond in a simple, instinctive way to signals that affect body functions essential for genetic fitness, i.e. survival and reproduction. These include the primitive sensations of pain (defined in this context as nociception), malaise (the sensation of feeling ill), hunger, and sex. All of these can be said to have an emotional content (feeling good or feeling bad) and can be a powerful motivating force for action; sex being the obvious example. I suggest that at this level the emotional responses to these sensations are entirely, or almost entirely instinctive and involve little, if any, thought. However, animals do not need to think about these sensations or express chronic anxiety at the possibility of their recurrence in order to experience distress when they occur. The property of sensation alone is sufficient to give animals protection under the UK Scientific Procedures Act (24). More on this later.

In almost all the animal species whose lives are affected by human contact, the expression of sentience is not limited to primitive sensation. The rules that govern our moral duty to respect their welfare must take account of biological evidence as to the extent to which they demonstrate the three inner circles of sentience, namely perception, mental formulation and consciousness. These are summarised in Table 11.1.

Animals that possess the property of perception do more than operate according to instinctive patterns of stimulus and response. They develop their minds. This involves

Table 11.1 Emotional and cognitive expressions of sentience with welfare implications

	Emotion	Cognition
Perception	Pain and fear	Avoidance
	Hunger and thirst	Food selection
	Comfort	Nest building
	Curiosity and security	Interpret simple social signals
Mental formulation	Anxiety and depression	
	Pleasure, joy, hope, grief	Recognition of social signals
Consciousness	Affiliative behaviour	Awareness of self and non-self
	Altruism and compassion	deceit

not only the recognition and identification of good and bad sensations but also the capacity to remember the association between physical and social messages leading to feelings such as pain, fear, hunger, comfort and security. This enables them to develop learned responses calculated to avoid future suffering and promote wellbeing. This increases their ability to mount an effective immediate response and improves the chances of doing things better next time. It also carries the potential to increase distress and anxiety if they learn that they cannot cope.

The properties of sentience described within (and limited to) circles two and three, sensation and perception, can be strictly linked to the genetic imperative to survive and reproduce, which means that they are entirely motivated by self-interest (or the interests of eight second cousins). Perception, defined here as a category of sentience, is more advanced than sensation because it involves learning, i.e. development of the mind. However, possession of (only) circles two and three does not admit any of the higher feelings or social graces. It has been relatively easy to demonstrate the property of perception, in its most primitive form, from simple laboratory experiments. Pavlov's classic conditioned response trials with dogs and bells, did no more than establish the link between perception of a conditioning stimulus (the bell) and salivation, the unconscious response to the arrival of food. Most of the animals I have discussed in this book can do better than this. Even Burns, for all his compassion for the startled mouse failed to grasp the nature of perception as defined here. *For thou art blessed, compared with me. The present only touches thee.'* Animals with the power of perception do not just live in the present. They remember, and this memory has long-term effects on their minds. If they learn to cope, they achieve an enhanced sense of wellbeing. If they discover that they cannot cope because they do not have the tools to cope or are so restricted that that they cannot make use of the tools at hand, they will suffer.

The fourth circle of sentience is defined as *mental formulation*: the ability to create mental pictures, or diagrams, that integrate and interpret complex information, experiences, sensations and emotions. Cognitive skills operating at this level include what behaviourists refer to as spatial awareness, an expression that rather belittles the amazing mapmaking skills of species like pigeons, horses, wildebeest and elephants. While much of the navigational equipment for long-distance migration in birds and marine animals is instinctive, the ability of the animals quoted above to carry within their minds detailed maps of areas extending to thousands of square kilometres exceeds anything you or I could manage. What these skills have in common is that they are skills that matter. They are critical to survival. In these migrant species, mental development has focused on mapmaking, probably to the exclusion of much else.

Possession of the capacity to construct complex mental formulations can have a profound impact on the emotional state and thereby the welfare of sentient animals in ways that can be good or bad. This property is particularly well developed (because it matters) in species that establish social relationships and form a society, like wolves, chimpanzees, and elephants, as distinct from animals like fish that just go around in large groups. Animals with the capacity to make complex mental formulations learn ways to distinguish between risks that are real and apparent, keep out of trouble, acquire an education, adapt to the demands of society, and thereby contribute to the mores of a culture. The emotional rewards accruing from this degree of mental development include, at least, a heightened sense of security and, at best, the pleasure to be gained from a life among friends. The downside of the property of mental formulation includes chronic feelings of anxiety or depression in animals that discover they are unable to cope with physical stresses like chronic pain or emotional stresses like bullying or isolation. It also includes grief at the loss of a friend or relative, although, at this level, it may have to be defined as a sense of personal loss, rather than compassion.

According to the skandhas, the word consciousness is restricted to properties within the final, deepest circle of sentience. As I wrote in Chapter 2, this is a difficult concept that can mean different things to different people. In psychology, it is used to describe the property of mind that can be pronounced most simply as the awareness of self and non-self. By this definition, conscious individuals are not only aware of the world around them but aware of the existence of themselves and others around them as distinct individuals with unique personalities, each with their own set of thoughts and feelings. The emotions made possible at this level of sentience include empathy, altruism and compassion, expressed, for example, in the form of affiliative behaviour: giving comfort and support to a neighbour in distress with no expectation of personal gain. This awareness of self and non-self (metarepresentation, or theory of mind, [22]) also has its selfish side. In human society, criminal fraud is based on getting into the minds of groups or individuals identified as vulnerable and bending their minds to your will.

These manifestations of the sense of awareness of self and non-self are thought by many to be unique to humans, although not necessarily all humans. It is argued that individuals with severe autism lack this property so are simply unaware of the feelings of others (51). The number of species for which we have good evidence for theory of mind is limited and has been largely restricted to social mammals e.g. great apes and dolphins, but the list is growing. It should probably include social corvids, e.g. rooks, (13) and possibly some invertebrates, e.g. squid and other cephalopods (69). Our duty to social species with these powers requires us to understand their need to communicate and respond appropriately to their social signals. However, I repeat what I wrote in Chapter 2. Animals do not need to display the deepest circle of sentience in order to suffer pain and distress, or experience comfort and joy. Moreover, they do not need to display the deepest circle of sentience to command our respect.

Outcome-based Ethics

I wrote at the outset, and I say it again: loving animals is not enough. If we are to do right by them, we need to understand them. On the other hand, loving animals can be too much. Our respect for the 'rights' of sentient animals has to be placed within the context of our own legitimate needs. This calls for some ethical decisions. For most of history, moral philosophy, or the study of ethics, has concerned itself exclusively with people. This anthropocentric tunnel vision was challenged by Albert Schweitzer, who wrote "*Ethics is nothing other than Reverence for Life. Reverence for Life affords me my fundamental principle of morality, namely, that good consists in maintaining, assisting and enhancing life, and to destroy, to harm or to hinder life is evil.*" (10)

If we are to adapt principles of ethics within human society into humane and effective action in respect to sentient animals, we need some ground rules. As with all sciences, ethics may be divided into the pure and the applied. Pure ethics is top-down. It asks the question 'Which general moral norms for the guidance of moral behaviour should we accept and why?' The aim of this approach is to justify moral norms. This is a worthy pursuit but one that may have little impact on the behaviour of society at large. Applied ethics is bottom-up, a more pragmatic business that begins with life as it is. It identifies a specific practical issue then constructs an analysis of relevant moral issues by a process of induction. Beauchamp and Childress proposed a practical, bottom-up approach to problems in medical ethics in the form of an 'Ethical Matrix' built on three aims of common morality, to promote *wellbeing, autonomy* and *justice* (2). This has been adapted by Mepham to consider issues relating to our treatment of the food animals (52).

Two key principles of ethics are Utilitarianism and Deontology. Utilitarianism is based on the principle of beneficence ('do good'), and non-maleficence ('do no harm') to promote the wellbeing of the greatest number. This principle has obvious relevance to our approach to animals other than pets because we have little option other than to consider them *en masse*. Utilitarianism has become a rather discredited concept in human philosophy because, considered in isolation, it neglects the rights of the individual. In the context of our contract with animals it isn't enough either. Philosophers describe our duty to respect individual rights as deontology, the science of duty, from the Greek *deon* (duty) and *logos* (science). It calls on us to respect the autonomy of all other individuals to those elements of wellbeing and freedom of choice that we might wish for ourselves and is most simply expressed as 'do as you would be done by'. If we are to apply the practical principles of bottom-up ethics, we cannot label ourselves as utilitarians or deontologists. We need to do both.

In recent years, it has become the custom to devise strategies for the promotion of health, safety and wellbeing in humans and other animals based on measures of outcomes rather than provisions: i.e. we measure the goodness of our actions by how well they work. This is a practical step in the right direction since outcomes are what matter. Outcome measures for the principles of utilitarianism and deontology are, respectively, the wellbeing of the population and the autonomy of each individual. The two outcome-based measures of wellbeing and autonomy as applied to all parties determine the third outcome-measure, *justice*. Justice demands that we, the moral agents, should seek a fair and humane compromise between what we take to meet our needs for food, sport, companionship, health and safety, scientific understanding, and what we give in terms of good husbandry, defined by competent, humane and sustainable action, for all life that comes within our dominion. This is not the same thing as giving all parties equal rights and will always fall short of the aspiration to seek total justice. It is an unavoidably distorted expression of Thomas Hobbes view of the Social Contract where individuals consent to surrender some of their freedoms and submit to an authority in exchange for protection of remaining rights and the maintenance of social order. In the early days of domestication, it may be argued that wolves consented to give up some of their freedoms in order to derive the benefits of living among humans. It would be ridiculous to suggest that the farmed animals consented to becoming food. However, it is, in my opinion, valid to argue that, on good farms with good husbandry, quality of life for the animals (while it lasts), when measured in terms of sustenance, comfort, companionship and security can be as good as, or even better than life in the wild.

Death and Killing

No discussion of ethical issues involved in our interactions with the other sentient animals can avoid the subject of death and killing. First, a disclaimer. I have killed animals, as humanely as possible, in the course of my work. When sailing, I have caught just enough mackerel for lunch. Since the age of ten, I have never killed a mammal, bird or fish for sport. At that age, on a friend's farm, I shot a sparrow with an air gun, watched it die, thought 'what a pointless waste of a life' and decided there and then never to do it again. My views on the killing of animals are inevitably personal and I accept that many will disagree (in both directions). All I can say is that I have been living with this topic for many years and what I write now is the best I can do to achieve some sort of compromise between my conscience and reality.

To begin with a few simple and incontestable statements of the obvious. Death is a fact of life for every living thing. Population control, by one means or another, is essential to the sustained life of the environment. The most worthy and organic gardener has to admit that 90% of their work involves the floral equivalent of butchery and slaughter. Being dead is not a welfare problem for the animal that has died, although it can be a problem for any offspring not yet mature enough for independent life. Unless death is immediate and totally unexpected, the act of dying does present problems of pain and fear. The claim that Shekita slaughter is humane because it is painless does not address the stress of fear in an animal conscious of choking to death in its own blood. The distress associated with dying and the approach of death will be a function of the intensity and duration of the period of suffering that precedes death. Whether the killer is a human, another animal, disease, or old age is irrelevant to the dying animal. A deer that gets a lethal shot from an expert stalker will suffer less than one hunted to exhaustion by humans, or wolves, and in the latter case, ripped to pieces. When a lion, that has depended on killing animals throughout adult life gets too old and toothless to hunt it is likely to suffer a prolonged decline before death from starvation or, when in extremis, being ripped to pieces by hyenas.

The more difficult question is 'do (non-human) sentient animals fear the prospect of death?' We really don't know. Sentient animals are clearly aware of danger when in proximity to a predator or in an area where they fear a predator may lurk. We may interpret this as fear of death, although we have no evidence that would enable us to conclude that they rationalise things this far. Because animals with the power of perception do not live only in the present, they remember sites and situations perceived to be dangerous, try to avoid them in the future, and where dependent offspring are involved, warn them against danger. Wild animals learn to live with threat and (another glimpse of the obvious) they only die once. Animals that perceive themselves to be under threat seek to maintain a safe flight distance from potential predators (whether humans or lions). Those forced to attempt an escape may be killed. Those that escape without harm remember that escape is possible so long as they are careful, so adapt to living under threat. Some years ago, I was examining records of fox hunting with hounds and discovered that the ratio of foxes roused by the hounds to foxes killed in the open field was thirteen to one. I think it reasonable to assume that the fox that escapes pursuit by predators develops the expectation that it can do it again so does not live in a state of constant fear. Animals that manage to escape and survive having been badly wounded will, of course, suffer.

The existential fear of death as the end of existence is, it would seem, part of human nature. Unless one believes in the fires of hell, this is an irrational fear since being dead is not a problem, although it does, of course, mess up one's long-term plans. Fear of suffering during the process of dying is real. I am with Woody Allen on this one: '*I have no fear of death. I just don't want to be there when it happens.*' Some animals, primates, cetaceans, elephants and rooks, appear to show signs of distress when faced by the death of a companion. This behaviour could reflect a sense of personal loss, compassion for a fallen comrade, or both, but I have no reason to suspect that their distress incorporates a sense of dread that at some undetermined time in the future this will happen to me.

Because the death of all animals is inevitable, whether or not humans are directly involved, I have no fundamental objection to the killing of animals for food or for population control as an instrument of environmental management when the alternatives can be worse. I am in total agreement with those who argue that humans eat too much food of animal origin for the good of our health and that of the environment. I try, but oft times fail, to abide by the principle 'Eat food, not too much, mostly plants'. I also abhor several approaches to the management of 'natural' habitats, such as the slaughter of raptor birds so that more grouse can be slaughtered by humans. Nevertheless, some natural environments such as the steppes of central Asia, the pampas of South America and the Taiga of the sub-arctic are sustained by stable populations of grazing animals. We cannot digest grass, and we need to control the animal population. In these circumstances, not eating the meat from these animals would be ecologically spendthrift. The killing of animals is necessary but because it is necessary; we have an obligation to ensure that the process of killing and everything that precedes that killing from the first moment of disruption to the normal routines of their life to the point of insensibility should be as quick and as humane as possible. This principle assumes the greatest importance in the case of the food animals, (from the farm or from the sea) because of the numbers involved and because of the potential to cause suffering during harvesting, transport and lairage in the abattoir. This is not the place to go into detail on welfare issues relating to the transport and slaughter of the food animals. Many, notably the Humane Slaughter Association (the sister charity of UFAW) have had much to say on this elsewhere. In the present context, the single message must be to strive to ensure that the animal to be killed passes from existence to non-existence with the absolute minimum of fear and distress. This principle should apply equally to the pig and to the lobster. Since death comes to us all, it may not be too much to suggest that this accords with the deontological principle of 'do as you would be done by'.

Killing of wild animals is an acceptable means of population control in order to preserve the balance of nature, when the alternatives are worse, both for nature and for individual animal welfare. About 30 years ago, as a member of the UK Farm Animal Welfare Council, I visited Oostvaardersplassen, established as a nature reserve on a reclaimed polder in Holland for birds that could come and go, and populations of horses, deer and cattle, who couldn't. The plan was to leave nature to itself to establish an equilibrium. We visited in March at the end of a hard winter when the grazing animals were dying of starvation in their hundreds. The cattle, constrained by the design of their mouths to harvest long grass, were the most affected. The deer and horses, more able to browse and tear out the remaining grass by its roots, took longer to die but were wrecking the pasture, shrubs and trees. I have no hesitation in stating that this was the worst systematic abuse of animal welfare I have ever seen. The tragedy was that it was done with the best of intentions. The practice continued for many years until the winter of 2016/7 when there were 3950 recorded deaths of cattle, deer and horses. Popular outcry finally forced a policy change towards population control through judicious culling.

Farms, Farmed Animals and Food

Table 11.2 applies the Ethical Matrix to the business of farming animals. In this model, the farmer is not just a food producer but also a steward of the living environment. The two moral agents in the matrix are, first, the farmers and landowners directly involved in the production of food from animals, and second, human society at large, which means all of us who derive any benefits from farmed land, whether for food, recreation or vital resources like water, soil and clean air: i.e. everybody. The moral patients are the farmed animals and the living environment. The right of the human population to demand food that is wholesome, safe and fairly priced carries the responsibility to all other concerned parties, farmers, farmed animals and the environment. In practical terms, it means that the public must accept the need for legislation to ensure acceptable standards of animal welfare and environmental care. I suggest that we also have a moral responsibility to go beyond the constraints of legislation to encourage incentives to improve husbandry (of animals and land). The most effective way to achieve this has been through consumer action, e.g. paying more (when we can) for food that carries the added value of local provenance, proven high welfare, and/or organic production methods.

Table 11.2 Food and farming: the ethical matrix.

	Wellbeing	Autonomy	Justice
Moral agents			
Producers and landowners	Financial reward Pride in work	Free competition	Fair trade Good husbandry Support for environmental schemes
Human society	Wholesome, safe, cheap food Access to the countryside	Freedom of choice	Added value for good husbandry
Moral patients			
Farmed animals	Competent and humane husbandry	Environmental enrichment Freedom of choice	'A life worth living'
The living environment	Conservation Sustainability	Biodiversity 'Live and let live'	Respect for environment and stewards

Justice for the producer balances the right to free trade and a decent income against the responsibility to practice good husbandry. According to the classic Adam Smith free-market argument, farmers should have no special rights. They are just one group within the overall division of labour so should be served no better or worse than any other group by the '*invisible hand of the market*'. If the sole function of the farmer were to provide food, this argument would apply. However, farmers also carry the direct responsibility to sustain the living environment, not just for us but for the sake of all life. This is a long-term commitment that cannot be addressed through the short-term economics of the free market. It is the responsibility of society at large. What this means in practice is that public money should be spent on public goods. Ideally, individual members of the public should meet the full cost of private goods, which implies that there should be no subsidies on food production *per se*. Individual buyers should be free to select the importance they give to price and production standards, which implies that those without money-worries should not have the right to impose high prices for value-added foods on those who cannot afford them. Stewardship of the environment (planet husbandry) is a public good so should be supported entirely from public funds (i.e. taxation). As I write, the UK government is examining ways to redirect agricultural subsidies towards the principle of public money for public goods. It remains to be seen how far they will go in this direction. I fear my proposal is probably too utopian to survive political debate. Nevertheless, it points in a good direction.

Justice for the moral patients requires us to provide the farmed animals with a life worth living. We can promote the wellbeing of the flock or herd through competent and humane husbandry. This may be expressed simply in terms of the five freedoms, freedom from hunger and thirst, chronic discomfort, pain, injury and disease, fear and stress and freedom of choice. The first four freedoms address the utilitarian principle of beneficence. Freedom of choice addresses our responsibility to recognise the autonomy of the individual by allowing the animals opportunities to make a constructive contribution to their own quality of life. This may involve selection of diet, environment (e.g. in or out of doors) and social expression. The utilitarian approach to the management of the living environment requires that we should strive to sustain the quality of the habitat and conserve all life within it. The principle of deontology identifies our responsibility to respect the individual needs of all fauna and flora, whether or not they make any contribution to our own wellbeing. We need to be realistic about this. We cannot and should not seek to preserve every rat and every tree. Nevertheless, we have a duty to respect the needs of all lifeforms, which implies that we should always seek to minimise individual suffering and lasting harm. This will require a policy of critical but constructive support to the farmers and landowners directly responsible for putting our principles of respect into practice.

Animals in Laboratories

Humans maintain and carry out experiments on animals in laboratories for the advance of knowledge through science, and for the protection of health and safety by testing a wide range of chemicals, such as medicines, garden sprays and common household products for evidence of toxicity. This is a utilitarian pursuit for the general wellbeing of the population of humans, domesticated animals and plants of nutritional or cultural interest. As with the farmed animals, the ethical issues are confounded by the fact that in the cost: benefit analysis all the benefits accrue to us, the moral agents and all the costs are borne by them, the moral patients. There has been much progress in recent years towards reducing the cost to animals of scientific procedures considered necessary for human health and safety. One big step in this direction has been adoption of the principle of the 'Three R's', reduction, replacement, and refinement. This calls on us to reduce the number of animals involved in experiments, where possible, replace animal-based tests with *in vitro* (test tube) experiments, and, when it is essential, to use live animals, refine the procedures to minimise the risk of incurring suffering. A big step forward was the UK was Animals (Scientific Procedures) Act 1986, which requires that procedures involving laboratory animals should only be allowed subject to a harm: benefit analysis where the cost to the animal in terms of 'pain, suffering, distress, or lasting harm' can be justified in terms of the potential benefit to humans (or other animals). The more recent EU Directive (2010/63/EU) carries a similar message. There can be no doubt that these directives have greatly improved conditions for laboratory animals, not least because they compel those who obtain their livelihood from working with these animals, whether directly or at a distance, to give more thought to how the animals might feel.

Table 11.3 uses the structure of the ethical matrix to examine the needs of the moral patients and the needs and responsibilities of the moral agents in the matter of procedures with laboratory animals. The first of the listed moral agents is human

Table 11.3 Application of the ethical matrix to procedures with laboratory animals

	Wellbeing	Autonomy	Justice
Moral agents			
Society at large	Health and safety	Freedom of choice	Compassionate and informed recognition of procedures
Regulators	Responsibility to society and experimental animals	Open minded approach to new developments	Welfare enshrined in codes of practice
Producers	Financial reward Responsibility to ensure welfare standards	Free competition	Compassionate interpretation of legislation Apply three Rs
Animal care staff	Pride in work Responsibility to animals	Freedom to make day-to-day decisions	Input into animal welfare policy
Moral patients			
Experimental animals	Minimal harm from procedures Day-to-day physical and mental wellbeing	Environmental enrichment Freedom of choice	Just application of harm: benefit analysis

society at large: i.e. all of us who depend on these procedures for our health and safety. We, as individuals, have the right to select from a range of products according to our individual perception of their efficacy and the possible cost of their use to laboratory animals and the environment. This freedom gives us the responsibility to develop a compassionate and informed understanding of what it is that we are buying. The regulators of scientific procedures have to balance their responsibility to society and to the experimental animals. The use of experimental animals for the testing of cosmetic products has been illegal since 1998 in the UK and 2009 in the EU. The principle of beneficence determines that animal welfare be enshrined in Codes of Practice. Autonomy, and the interests of animal welfare, require the regulators to be open-minded and sympathetic to new ideas. This is particularly relevant to the business of routine toxicology testing, where it is too easy to stick with old routines when more refined, less stressful alternatives become available because 'we have always done it this way'.

Developers and producers of new drugs, household goods and garden sprays have the right to financial rewards for their work and to free competition, unimpeded by restrictive legislation. In return, they have the responsibility to promote the welfare of both the experimental animals and the staff with direct responsibility for their care. The animal care staff in scientific laboratories carry out physically and emotionally taxing work with a great deal of skill and compassion. It is the responsibility of their bosses to ensure not only that they can take pride in their work but that their voices are heard in regard to day-to-day and strategic decisions in relation to matters of animal husbandry and welfare.

Our first duty to the animals is to minimise harms arising from the procedures themselves through just application of the harm:benefit analysis. In the UK, this is enshrined in law and effectively policed through formal inspections and the presence of a named veterinarian whose first responsibility is to the animals, not the procedures. However, seeking to minimise harm during the procedures is not enough. Most animals in laboratories, most of the time, are not actually undergoing potentially harmful procedures, but may be living a depressing life in barren cages. When not constrained by the needs of a specific procedure all these animals should be given the opportunity for as much environmental and social enrichment as possible.

Generally speaking, laboratory animals receive more protection from the law than farmed animals and are less likely to suffer neglect. It is, moreover, instructive to compare the relative magnitude of welfare problems for farmed and laboratory animals. To paraphrase a paragraph in my earlier book, Animal Welfare: A Cool Eye towards Eden, '*In the UK, the average human omnivore who maintains a good appetite to the age of seventy will consume 550 poultry, 36 pigs, 36 sheep and six oxen. The number of mice sacrificed to advance knowledge and improve human health and safety is two.*' I suggested at the time (1994) that this may not be too much to ask of brother mouse. The most recent statistics from the Home Office (2016) show that this number has not significantly changed.

Wild Animals in Captivity

I repeat: our best policy in regard to animals in the wild is to leave them well and leave them alone. To achieve a sense of wellbeing, they need an environment that provides the full range of the resources they require to meet their physical and emotional needs in all seasons. For the larger mammals, this implies a varied habitat and a lot of space. Currently, this space is being eroded at a rate that poses an existential threat to many species. The most effective way to address this problem is through the establishment of large nature reserves. When this is not possible, smaller reserves can be strategically linked by wildlife corridors that allow free movement of animals between them. Since it is human nature to invade and erode these areas of wilderness for our own devices, it will nearly always be necessary to manage and police these reserves to prevent poaching and destruction of habitat. This costs time and money, so it makes complete sense to provide financial support from humane, environmentally sustainable tourism. Management of even the most extensive of these reserves is likely, on occasions, to require a controlled cull of iconic but environmentally destructive species like elephants in order to avoid a catastrophic failure of good intentions such as the Oostvaardersplassen fiasco.

The aim of nature reserves is to sustain a habitat wherein wild animals adapted to that habitat have the freedom to live the life for which they are best suited and, if these are managed well, this would seem to be beyond reproach. More contentious is the business of keeping wild animals in captivity for the enlightenment and entertainment of the paying public. When I was a small boy living in Bristol, I greatly enjoyed trips to the zoo to get a ride on the elephant and gaze in wonder at large cats (and a famous gorilla) in cages. In common with almost everybody else at the time, I gave little thought as to how the animals might feel about life in these conditions and I was totally unaware that most of them had been captured from the wild and locked up for life. Thankfully, zoos like this are no longer tolerated in most societies. Much thought has been given to the creation of enriched environments that seek to reproduce the resources of the natural habitat within the confines of available space and the need to give the public the chance to see the animals. The original site of Bristol Zoo will close in 2022 and a new, more extensive and natural 'zoo' will open in 2024. Most zoo animals are now bred in captivity or, when at threat of extinction in the wild, brought to zoos and wildlife parks for breeding to preserve the species and, in a few cases, for return to the wild.

I shall not dwell on the motivation of those who run zoos for the public, not least because this book is about animals, not people. I would just say that my experience of those who work in zoos at all levels is that they love their animals and most of them understand them pretty well. In keeping with my general theme, my question is 'How do 'wild' animals feel about life in captivity?' I put the word 'wild' in quotation marks because the phrase wild animal, like farm animal, is an anthropocentric conceit; classification of animals in terms of their utility to us. When we are unsure about how to act in relation to an animal of which we know little, it is best to start from first principles. Any sentient animal, however, we may choose to categorise it, is driven by the motivation to satisfy its physical and emotional needs and seek the freedom to control its own quality of life. Basic needs include freedom from hunger and thirst, pain, fear, injury and disease, and freedom to seek comfort and security. Advanced psychological needs at the deeper level of sentience may include freedom to express curiosity and pleasure, play and have a good social life. The relative importance of these different needs will differ between different species adapted over generations to domestication or life in the wild. However, these principles can be applied to any set of questions relating to the basic and advanced needs of sentient animals wherever they may be, addressed to those with direct responsibility for control of their lives and answered on a species-by-species basis by those with the competence to judge. I illustrate this principle with two examples.

The need to eat is the strongest of the life forces and one that cannot be measured simply in terms of the physical need to maintain normal body function through the provision of adequate nutrition. Sentient animals have a strong emotional need to forage or hunt for food and derive great satisfaction when that need is met. Good zookeepers know well that their animals derive satisfaction from being made to work for their food reward and will go as far as reasonably practical to make the hunt as realistic as possible. Obviously, offering live goats to tigers is not on, although the ban on live prey is not universal. Some snakes may only eat if presented with a live mouse. With herbivores, the situation is rather different. Equines, by virtue of the design of their digestive tract have a physical need to forage for many hours. Elephants in the wild, by virtue of their large appetite for foods of low digestibility need to travel for long distances to meet their nutrient requirements. Some argue from this observation of natural behaviour that elephants have a basic need to take long walks in wide open spaces. Possibly they do, but I suggest that it is more likely that they are motivated to spend many hours foraging for food, rather than simply going for a long walk. With both carnivores and herbivores, the aim should be to make the search for food a pleasant and rewarding way of passing the time.

Meeting the needs of animals in captivity for a secure and satisfactory social life is a critical issue that must be addressed on a species-by-species basis and by those with the competence and experience to speak for the species in question. Some crave social contact and suffer in isolation. Others are adapted to a solitary existence. Still others can elect to form close relationships with individuals not necessarily of the same species but remain permanently antagonistic to others. Some animals experience fear and stress when humans are in sight, and they perceive they have no place to hide. Others clearly enjoy human company. In captivity, these are likely to be learned choices rather than a hard-wired property of the species. I have seen and photographed a young rhino in the extensive paddocks of the Longleat Safari park approach one of the animal attendants (a veterinary student with a summer job) to get a tickle under the chin. This could hardly be called 'natural' rhino behaviour but, given freedom of choice, this young rhino elected to seek out something that it had learnt would give it pleasure.

Animals in Sport and Entertainment

Throughout history, animals have been used by humans for a wide range of entertainments, most of them barbaric (83). Lethal pursuits included the baiting of bulls and bears, bull fights, dog fights, cock fights and eating Christians. Non-lethal attractions for our titillation have involved performing animals, especially in circuses. The animal most commonly required to perform for humans in the pursuit of sport and recreation is the horse, exploited for its athletic prowess and adaptability to human training methods. Performances that reflect agricultural practices and traditions include rodeo and sheepdog trials. I exclude from this discussion any consideration as to the morality of legal sporting pursuits that involve the killing of animals (hunting, shooting, fishing) on the grounds that this book is an enquiry into the minds of sentient animals and no sentient animal in its right mind would voluntarily wish to take part.

The three questions that need to be addressed are:

- Can the performance, whether in the circus or on the racetrack be a source of suffering?
- Do the training methods deemed necessary to prepare the animals for the performance involve the deliberate imposition of pain or fear?
- Do the living conditions of the animals meet their needs for the times when they are not performing? (which is, of course, most of the time.)

Horse racing exploits the strong motivation of the horse to run, and to run in company. In nature, and when not driven by jockeys, some horses prefer to lead, others to follow. In general, I believe that horse racing, whether on the flat or over fences, is an activity that depends on the willing, indeed enthusiastic participation, of the horses. Horses do, of course, get serious and lethal injuries on the racetrack and these are of great concern to owners, carers and the general public. Those with the most severe injuries are likely to be killed on the track, so they don't suffer long. Those with less severe injuries that are allowed to recover may never race again so will not experience the fear of its recurrence. In this regard, racing differs from eventing when horses are expected to clear extremely challenging jumps, frequently come to grief but are expected to go through the same routine again and again. I heard a famous event rider talk of 'Badminton virgins': young enthusiastic horses with, at that stage of their lives, no experience of falls and therefore no fear. He also spoke of experienced horses that learned to be more circumspect and of occasions when he has pulled up a horse because it was clearly hating every moment of the ride. This very experienced and sympathetic rider was making the point that eventing can become distressing to some horses.

The biggest point of contention in regard to the welfare of racehorses relates to the use of the whip. This is, by definition, based on negative reinforcement, the stimulus of pain (or fear of pain when the jockey shows the horse the whip) to drive the animal harder. There are strict rules on the design and use of the whip intended to minimise stress, but it could never be construed as other than a violent act carried out exclusively for the benefit of humans involved (jockey, owner and trainer). There is now pressure to ban the whip altogether. Before this proposal is dismissed as sentimental nonsense, we should consider, first, whether a whipped horse actually runs faster, and the evidence for this is shaky; secondly whether racing without whips would make any difference to the excitement generated by the race and the attractions of betting. Horses are going to finish the race in an order determined by their fitness, their motivation and the skill of their jockey. This will happen whether or not the jockeys carry whips, and the bookies and punters will, on average, be neither better nor worse off. Owners may grumble.

Paradoxically, the equestrian event most open to informed criticism on welfare grounds is the one that involves the least effort and the least risk, namely dressage (48). The horses are trained to perform a number of unnatural manoeuvres while their head and neck are forced into an unnatural posture. To achieve this unnatural posture many dressage horses are compelled to be held on a tight rein pulling on bridles with an extremely tight nose band. This forces the horse to adopt the unnatural head position because it is painful to do otherwise. Moreover, there is evidence that the prolonged use of tight nose bands can lead to lasting damage to the mouth. Once again, the traditionalists argue that dressage, as currently practised, is a supreme demonstration of the ability of the horse to perform a complex series of manoeuvres devised by us. I suggest (and I am not alone) that it would be an even more impressive demonstration of the skills of horse and rider if it could be achieved without recourse to bit and bridle.

The circus has been the traditional place to watch performing animals. Once again, in my boyhood days, this involved not only horses, but lions, tigers and seals. Today, the use of animals such as the large cats, which had to be trained by methods based on fear, is (entirely?) extinct. The few animal acts we see in the circus today usually involve animals like dogs and horses that can be trained by positive reinforcement. When these animals are trained well and treated well, I do not think they present a welfare problem. Some will argue that dressing animals up in silly costumes is an insult to their dignity. If we stick strictly to our understanding of animal behaviour, this will only present a problem for animals that possess a sense of self and non-self, and therefore an awareness of what they look like. We may exclude horses from this category. There is some experimental evidence that dogs have a sense of self and non-self so could be distressed by the fact that their appearance is unnatural. It is much more likely that dogs pick up human signals that cause them distress because they sense they are being laughed at. They will obviously be irritated if the article of apparel itches or interferes with their behaviour. I think it entirely proper to object to human actions that abuse the dignity (the telos) of our fellow mortals on the grounds that they reflect some of our most extreme displays of sentimentality and anthropomorphism. However, I say again, so far as the animals are concerned, it is not what we think but what we do that matters. It follows that I am not too concerned by performing animals acts that cause no distress to the animals, either in performance or in training. My main concern as to the welfare of animals in circuses (and bad zoos) relates to question three. The quality of their living accommodation can fall far short of that necessary to meet the five freedoms.

I include in my list of animal performances rodeo and sheepdog trials. Many of the events in a western rodeo are designed to show off skills necessary to the work of the traditional cowboy but two of the most spectacular events, bronco riding and bull riding, are completely pointless. The rider has to spur the bucking animal and stay on board for eight seconds. Equal points for performance are awarded to animal and rider. This looks like a violent and cruel practice, although the only animal likely to be injured is the cowboy. Aversion techniques are used in training animals for the rodeo, but once trained, they become seasoned performers. Moreover, the contest, as seen through their eyes, is one they always win, since after no more than eight seconds, the cowboy always comes off. I have spent some time observing the behaviour of experienced rodeo horses and bulls and, generally speaking, out of the ring, they appear to be relaxed and stress-free. Compare this with the experience of sheep in that most serene and British of pastimes, the sheepdog trial. When dog and handler are expert, the sheep will be moved with minimal distress at a controlled speed and in the right direction because all parties, sheep, sheepdog and shepherd, understanding the rules of flight distance. When things go wrong, the sheep run off, maybe not in panic, but under stress. In any event, it is a game that the sheep will always lose. I am a fan of sheepdog trials because I believe they are a beautiful display of human and dog working in combination to perform a task essential to the humane and sensitive management of an only slightly domesticated animal. However, when viewed through the eyes of the animals I can't see much wrong with much of rodeo either. There is one exception. The only rodeo practice that has any real relevance to the practical skills needed by the cowboy is the one likely to cause distress, namely calf roping: lassoing a young calf from horseback.

Pets

Most of the big issues arising from our interactions with the animals we love the most were covered in Chapter 10. In this final chapter, it is worth revisiting some of these issues because our pets (excepting cats) are the animals whose lives are most affected by our behaviour. Dogs that are almost entirely dependent on us are potentially most at risk from physical abuse, such as starvation and neglect (which may be rare) and emotional stress, such as separation anxiety (which is all too common). Horses designed to graze the open plains in a herd of close compatriots can suffer physical and emotional stress as a result of improper feeding and social isolation. Breeding animals to suit the whims of fashion abuse our responsibility to ensure the fitness of animals in our care. For reasons of fashion, or perceived cuteness, we breed dogs that are prone and, in some cases, condemned to suffer from conditions that are entirely of our own making. These range from chronic respiratory distress to the need for repeat Caesarian sections. Even horses are now being bred for traits that may look pretty (or fashionable) to our eyes but impair their normal athleticism. I cite the breeding of Arab horses with noses dished to a degree that significantly impairs respiration, especially during exercise.

We expect some animals classified by us as pet or companion to do more that conforms to a lifestyle of our choosing. We also expect them to do our will. This applies most obviously to dogs and horses. When this is done well it can be very satisfactory for both parties, especially in the case of dogs whose minds are naturally inclined to seek the approval of their leaders and make a positive contribution to the welfare of the extended family. Well-trained contributors to the general good, such as police dogs, guide dogs, and sniffer dogs undoubtedly derive satisfaction from their work and live a more fulfilling life than those shut up indoors with nothing to do all day. Well-trained dogs and horses are emotionally stable, and their quality of life is likely to be better than just satisfactory. Expert trainers, consciously or otherwise, understand the minds of their animals. The sentient minds of non-human animals are complex but not complicated. They recognise direct, consistent, unambiguous signals and learn to react as we would wish when correct responses are rewarded by positive reinforcement. Domesticated animals cannot be expected to do our will if they fail to understand the message, the messages are inconsistent, or they are punished for doing what they thought to be the right thing.

I recognise that most of the people who insult the natural integrity or confound the minds of their pet animals are unaware that they are doing any harm. Should any of them have read this book and are anxious that they may fall within this category, I would ask them to follow the simple message that applies to all who have responsibility for the care of sentient animals, whether in the home or on the farm, in laboratories or zoos. Get as much good advice as you can from those with real experience, then work from first principles. Lay out the set of basic and advanced physical and emotional needs that apply to all sentient animals. Here, the five freedoms can provide a sturdy framework on which to hang the details. Apply these basic principles to meet the special needs of the animals for whom you have responsibility within the environment you have imposed on them. Regularly review the outcome of your actions in the context of how well the physical and emotional state of your animals appears to meet these five freedoms and, if necessary, modify your own actions. Your aim must be to ensure, so far as possible, that you and the animals in your care are of the same mind.

What can We Learn from the Animals?

In my final brief summary of a complex story, I shall try to follow the advice of Einstein, (who managed some very complex ideas). He said, '*Everything should be kept as simple as possible – but no simpler.*' My first simple, but not simplistic, assertion is that our similarities with other sentient species far outweigh the differences. Within all but the deepest circle of sentience (consciousness), our own sensations, perception and motivation to behaviour operate according to the same drives that determine the behaviour of the other sentient animals. Our divided brains interpret incoming sensations and information that matter in two ways. Keeping things very simple indeed (too simple for some) we can picture a brain in which the right side gives us the big picture, whose impact is measured emotionally. Does this make me feel good, bad, or indifferent? The left side reviews the details and helps us to make a cognitive, considered response. In nearly all cases, however, this cognitive response is directed to fine-tune the way we feel. In most circumstances, the extent to which human behaviour is governed by knowledge and understanding is likely to be greater than in other animals because we can call on a much greater reservoir of knowledge, made available largely through the power of language and our ability to conserve and communicate language-based knowledge through books and other media. This can help us to do sensible (rather than impulsive) things. It can also lead us to engage in self-indulgent, post-hoc rationalisation designed to justify behaviour that was, at the time, dictated almost entirely by emotion and personal satisfaction.

All sentient animals are motivated by self-interest. There is a grain of sense in the mad assertion that it is the aim of every animal species to convert all organic matter into itself. Fortunately, in a sustainable environment, no single species has the ability to achieve this. Each may be out for itself but, in the absence of human interference, its capacity to do serious environmental harm is self-limiting. Humans, given our greater store of knowledge and our capacity to call on forces stronger than ourselves, such as fossil fuels and high explosives, have a far greater capacity to do real and lasting harm and we have been far too indulgent in this capacity. We may argue in our defence that we differ from other animals in that we have sufficient depth of consciousness and powers of moral judgement to develop ethical codes and social practices designed to deliver us from lives that are 'solitary, nasty, brutal and short'. It would, however, be unrealistic in the extreme to assert that this is an innate, universal property of human behaviour. We cannot escape the fact that much human behaviour, including apparent altruism, is likely to be dictated by the primitive need to feel good about ourselves. Indeed, this is the basis for one popular psychological, if unscientific, explanation of human motivation, Maslow's Hierarchy of Needs (44). This can be illustrated by a pyramid with basic physiological needs for food, water, safety and security at the base, psychological needs for belonging and love (the basis of affiliative behaviour) in the middle and, at the apex, the need for self-fulfilment – (achieving one's potential, creative activity). Figure 11.1 shows dotted arrows pointing to the two most basic needs, metabolism, safety and security – homeostasis in regard to both the internal and external environments. In humans, as in all animals with sentient minds (as distinct from sensation alone), all needs have a psychological component.

Maslow's hierarchy of human needs appears to place us fairly and squarely within the world of sentient animals who operate quite successfully without moral codes. Where it differs, of course, is that it includes the motivation to self-fulfilment. In this respect, we may be able to class the human species as unique. I recall the words of Walt Whitman with which I prefaced my first section on sentience and sentient minds '*I think I could turn and live with animals. I look at them long and long. They do not lie awake at night and whine about their sins. Not one is dissatisfied. Not one is demented by the mania of owning things. Not one is respectable or unhappy over the whole earth.* Whitman is clearly wrong in his assertion that no animal is dissatisfied. This can be explained by the fact that he was observing animals with a great sense of empathy but through his eyes, not theirs. However, I entirely subscribe to his belief that non-human animals are happily free of any anguish associated with the drive to self-fulfilment.

I cannot accept Maslow's hierarchy, even as an explanation for human motivation because I believe it is, fortunately, impossible to reject the belief that while most human behaviour is primarily motivated by self-interest, our species includes that aspect of deep consciousness that we describe as morality. I turn again to the words of Albert Schweitzer. '*The most immediate fact of man's consciousness is the assertion "I am life that wills to live in the midst of life that wills to live*' (10). What he is saying in plain words, is that we differ not at all in our primary motivation from all the other sentient animals. He is clear, however, that we have the moral capacity to do better than this.

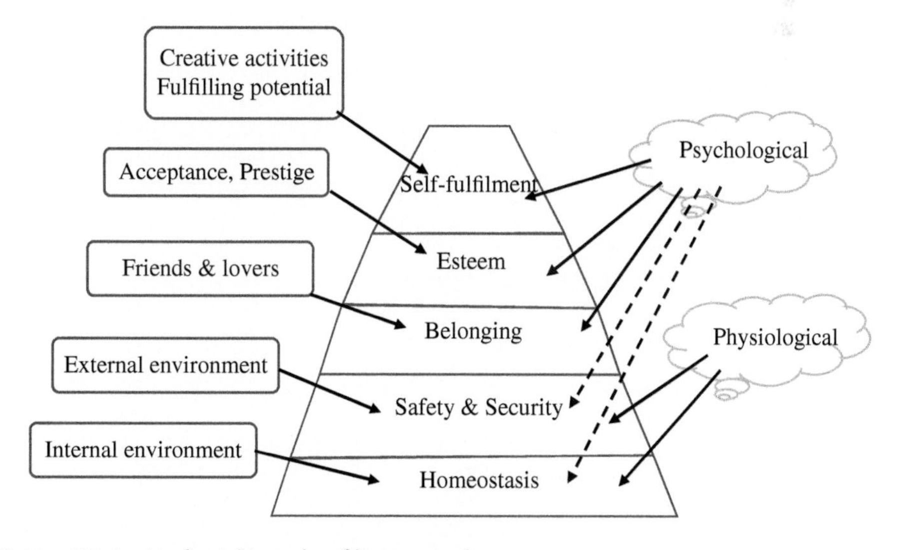

Figure 11.1 Maslow's hierarchy of human needs

'The essential element in civilization is the ethical perfecting of the individual as well as society. At the same time, every spiritual and every material step forward has significance for civilization. The will to civilization is, then, the universal will to progress that is conscious of the ethical as the highest value'. The ethical perfecting of the individual as well as society is a (perfectionist) outcome measure of the two moral principles of deontology and utilitarianism, where the terms individual and society extend to all living forms. It is, in a nutshell, the principle of reverence for life.

To say that humans are moral creatures is not the same as to say we are all good people. Morality, like welfare, operates within a spectrum from the very good to the very bad. Morality is a consequence of the biological faculty of metarepresentation that makes us aware of the circumstances, motivation and behaviour of others and takes these things into account when we make our decisions. At best, this can lead us to acts of charity and self-sacrifice that go far beyond that which any other species could imagine. This is just as well since it has to be set against our capacity to exploit the wealth and resources of the living planet for our own selfish ends and to hell with the rest. This aggressively self-serving behaviour, which also goes far beyond that which any other species could imagine, has undoubtedly favoured certain individuals and certain populations within human society but, because Darwin is always right, it will, in the long term, not succeed because it is ecologically non-adaptive. If we are all to thrive and sustain a sense of wellbeing, we will need to focus on the goal of civilization, absolutely not in the Platonic sense of utopian city states but as used by Schweitzer in context of the principle of reverence for life. I can do no better than to close with the words of Schweitzer as paraphrased by James Brabazon (10), *'We are brothers and sisters to all living things and owe to all of them the same care and respect that we wish for ourselves.'*

Further Reading

Specific references in text (Some of these also appear under General Reading)

1. Baron-Cohen, S., Leslie, H.M., and Frith, U. (1985). Does the autistic child have a theory of mind? *Cognition* 21: 37–46.
2. Beacham, T.L. and Childress, J.F. (1994). *Principles of Biomedical Ethics,* 4e. New York: Oxford University Press.
3. Bearzi, G., Kerem, D., Furey, N.B. et al. (2018). Whale and dolphin behavioural response to dead conspecifics. *Zoology* 128: 1–15.
4. Bell, F.R. and Sly, J. (1979). The metabolic effects of sodium depletion in calves on salt appetite as assessed by operant methods. *Journal of Physiology* 195: 431–442.
5. Benz-Schwarzburg, J. and Knight, A. (2011). Cognitive relations yet moral strangers? *Journal of Animal Ethics* 1: 9–36.
6. Berdoy, M. (2003). The Laboratory Rat: A Natural History. https://www.youtube.com/watch?v=giu5WjUt2GA (accessed 12 October 2021).
7. Biro, D., Freeman, R., Meade, J. et al. (2007). *P*igeons combine compass and landmark signals *PNAS* 104: 7141–7476.
8. Bizzozzero, M.R., Allen, S.J., Gerber, L. et al. (2019). Tool use and social homophily among male bottlenose dolphins. *The Royal Society Publications* 286: 1–22.
9. Boesch, G., Hohmann, G., and Marchant, L. (2002). *Behavioural diversity in chimpanzees and bonobos.* Cambridge: Cambridge University Press.
10. Brabazon, J. (1975). *Albert Schweitzer: A Biography.* Syracuse University Press.
11. Brown, R. (1958). *Words and things. An Introduction to Language.* New York: Free Press.
12. Brunton, C.F.A., Macdonald, D.W., and Buckle, A.P. (1993). Behavioural resistance towards poison baits in brown rats *Rattus norvegicus. Applied Animal Behaviour Science* 38: 159–174.

Animal Welfare: Understanding Sentient Minds and Why it Matters, First Edition. John Webster.

13. Clayton, N.S. and Emery, N.J. (2007). The social life of corvids. *Current Biology* 17(16): R652–656.
14. Crook, R.J. (2021). Behavioural and neurophysiological evidence suggests affective pain experience in octopus *iScience* 24(3): 102229.
15. Clutton-Brock, T.H., Hiraiwa-Hasegawa, M., and Robertson, A. (1989). Mate choice on fallow deer leks. *Nature* 340: 463–465.
16. Dawkins, M. (1993). Through our eyes only?: The search for animal consciousness. Oxford: Freeman.
17. Dixon, R.M., Fletcher, M.T., Goodwin, K.L. et al. (2017). Learned behaviours lead to bone ingestion by phosphorus-deficient cattle. *Animal Production Science* 59: 921–932.
18. Douglas-Hamilton, I., Bhalla, S., Wittemyer, G., and Vollrath, F. (2006). Behavioural reactions of elephants towards a dying and deceased matriarch. *Applied Animal Behaviour Science* 108: 87–102.
19. Evans, A.L., Singh, N.J., Friebe, A. et al. (2016). Drivers of hibernation in the brown bear. *Frontiers in Zoology* 13: 7.
20. Freire, R., Wilkins, L.J., Short, F., and Nicol, C.J. (2003). Behaviour and welfare of individual laying hens in a non-cage system. *British Poultry Science* 44: 22–29.
21. Frith, C. and Frith, U. (2005). Theory of mind. *Current Biology* 15: 644–645.
22. Gallup, G.G., Anderson, J.R., and Sillito D.J. (2000). The mirror test in *The Cognitive Animal* https://courses.washington.edu/ccab/Gallup%20on%20mirror%20test.pdf
23. Gestalt psychology https://en.wikipedia.org/wiki/Gestalt_psychology
24. Government Legislation (1986). Scientific Procedures Act (UK). www.legislation.gov.uk/ukpga/1986/14/contents (accessed 14 October 2021).
25. Government legislation (2006). Animal Welfare Act (UK). www.legislation.gov.uk/ukpga/2006/45/contents (accessed 12 October 2021).
26. Griffin, D.R., Webster, F.A., and Michael, C.R. (1960). The echolocation of flying insects by bats. *Animal Behaviour* 8: 141–154.
27. Gruber, T. (2016). Great apes do not learn novel tool use easily: conservatism, functional fixedness or cultural influence? *International Journal of Primatology* 37: 296–316.
28. Gruter, C. and Farina, W.M. (2009). The honeybees waggle dance: Can we follow the steps? *Trends in Ecology and Evolution* 24: 242–247.
29. Harrison, R. (2013). *Animal Machines. The New Factory Farming Industry*. Oxford: CAB International.
30. Hume, D. (1740). Treatise on Human Nature, Book II. files.libertyfund.org>files (accessed 14 October 2021).
31. Janik, V.N. and Sayigh, L.S. (2013). Communication in bottlenose dolphins: 50 years of signature whistle research. *Journal of Comparative Physiology* 199: 479–499.
32. Johnsen, S. and Lohmann, K.J. (2005). The physics and neurobiology of magnetoreception. *Nature Reviews Neuroscience* 6: 703–712.
33. Kendrick, K.M. (1991). How the sheep's brain controls the visual recognition on animals and humans. *Journal of Animal Science* 69: 5008–5016.
34. Kestin, S.C., Gordon, S., Su, G., and Sorensen, P. (2001). Relationships in broiler chickens between lameness, liveweight, growth rate and age. *Veterinary Record* 148: 195.
35. Keverne, E.B. and Kendrick, K.M. (1992). Oxytocin facilitation of maternal behaviour in sheep. *Annals New York Academy of Science* 652: 83–101.
36. Kipling, R. (1902). *Just So Stories*. London: Macmillan.
37. Kirkwood, J.K. and Webster, A.J.F. (1984). Energy budget strategies for growth in mammals and birds. *Animal Production* 38: 147–156.

38. Kirkwood, J.K. (1985). Patterns of growth in primates *Journal of Zoology* 205: 123–136.
39. Knowles, T., Kestin, S.C., Haslam, S.M. et al. (2008). Leg disorders in broiler chickens: prevalence, risk factors and prevention. *PLOS One* 3: e1545.
40. Krupenye, C. and Call, J. (2019). Theory of Mind in Animals: Current and future directions. *WIREs Cognition Science* 10: 76–87.
41. Kyriazakis, I. and Oldham, J.D. (1993). Diet selection in sheep: the ability of growing lambs to select a diet that meets their crude protein requirements. *British Journal of Nutrition* 69: 617–629.
42. Leavens, D.A., Russell, J.L., and Hopkins, W.D. (2009). Multimodal communication by captive chimps (*Pan troglodytes*). *Animal Cognition* 13: 33–40.
43. Lemon, R. (1975). How birds develop song dialects. *The Condor* 77: 383–400.
44. Maslow, A. (1954). *Motivation and personality*. New York: Harper.
45. Mason, G.J., MacFarland, D., and Garner, J. (1998). A demanding task: Using economic techniques to examine animal priorities. *Animal Behaviour* 55: 1070.
46. McFarland, D. (1989). *Problems of Animal Behaviour*. Harlow: Longman.
47. McGreevy, P.D., Webster, A.J.F., and Nicol, C.J. (2001). Study of the behaviour, digestive efficiency and gut transit times of crib biting horses. *Veterinary Record* 148: 592–596.
48. McLean, A. and McGreevy, P. (2010). Horse training techniques that may defy the principles of learning therapy and compromise welfare. *Journal of Veterinary Behaviour* 5: 187.
49. McCleod, S. (2018). Tolman-Latent learning. www.simplypsychology.org (accessed 14 October 2021).
50. McGilchrist, I. (2012). *The Master and his Emissary. The Divided Brain and the Making of the Western World*. New Haven: Yale University Press.
51. Mendl, M., Burman, O.H.P., Parker, M.A., and Paul, E.S. (2009). Cognitive bias as an indication of animal emotion and welfare. Emerging evidence and underlying mechanisms. *Applied Animal Behaviour Science* 118: 161–175.
52. Mepham, B. (1996). Ethical analysis of food technologies: an evaluative framework. In: *Food Ethics* (ed. B. Mepham), 101–119. Routledge.
53. Mowatt, F. (1996). *Never Cry Wolf The amazing true story of life among arctic wolves*. Boston: Little, Brown and Company.
54. Nicol, C.J. and Pope, S.J. (1996). The maternal feeding behaviour of hens is sensitive to perceived chick error. *Animal Behaviour* 52: 767–774.
55. Olds, J. (1958). Self-stimulation of the brain. *Science* 127: 315–324.
56. Paulos, R.D., Trone, M. and Kuczaj, S.A. (2010). Play in wild and captive cetaceans *International Journal of Comparative Physiology* 23: 701–722.
57. Pepperberg, I.M. (2006). Cognitive and communicative abilities of grey parrots. *Applied Animal Behaviour Science* 100: 77–86.
58. Pfister, J.A., Muller-Schwartz, D., and Balph, D.F. (1990). Effects of predator fecal odors on feed selection by sheep and cattle. *Journal of Chemical Ecology* 16: 573–583.
59. Protection of Animals Act 1911 https://www.legislation.gov.uk/ukpga/Geo5/1-2/27/contents
60. Raihani, N.J. and Ridley, A.R. (2008). Experimental evidence for teaching in wild Pied Babblers. *Animal Behaviour* 75: 3–11.
61. Rogers, L. (1980). Lateralisation in the avian brain. *Bird behaviour* 2: 1–12.
62. Rogers, L. (2006). Cognitive and social advantages of a lateralised brain. In: *Behavioural and Morphological Asymmetries in Vertebrates* (ed. Y.B. Malashichev and A.W. Deckal), 129–139. Boca Raton: CRC Press.
63. Rupert, A., Moushegian, G., and Galambos, R. (1963). Unit responses to sound in the auditory nerve of the cat. *Journal of Neurophysiology* 26: 449.

64. Sayigh, L.S., Carter Esch, H., Well, R.S., and Janik, V.M. (2007). Facts about signature whistles of bottle-nosed dolphins. *Tursiops truncates Animal Behaviour* 74: 1631–1642.
65. Skandha in Buddhist teaching https://en.wikipedia.org/wiki/Skandha
66. Sherwin, C.M. and Nicol, C.J. (1997). Behavioural Demand Functions of Caged Laboratory Mice for Additional Space. *Animal Behaviour* 53: 67.
67. Silk, J.B. (2001). Empathy, sympathy and prosocial preferences in primates. In: *Oxford Handbook of Evolutionary Psychology* (ed. L. Barrett), 115–126. Oxford: Oxford University Press.
68. Smith, B. (1988). *Foundations of Gestalt Theory* https://philpapers.org/archive/SMIFOG.pdf
69. Smith, G. (2020). *Metazoa: Animal Minds and the birth of consciousness*. London: William Collins.
70. Smith, P.A. and Edwards, D.B. (2018). Deceptive nest defence in ground nesting birds and the risk of intermediate strategies. *PLOS One* 13(10): e0205236.
71. Sneddon, L.U. (2009). Pain perception in fish: indicators and endpoints. *ILAR Journal* 50: 338–342.
72. Sneddon, L.U., Braithwaite, V.A., and Gentle, M. (2003). Novel object test: examining nociception and fear in the rainbow trout *The Journal of Pain* 4: 431–441.
73. Treaty of Amsterdam 1997 The Lisbon treaty: recognising animal sentience https://www.ciwf.org.uk/news/2009/12/the-lisbon-treaty-recognising-animal-sentience
74. Universities Federation for Animal Welfare (UFAW) *Guiding principles for the humane control of rats and mice*. https://www.ufaw.org.uk/downloads/welfare-downloads/guidance-on-humane-control-of-rodents-feb2509v19.pdf
75. Villanueva, R., Perricone, V., and Fiorito, G. (2017). Cephalopods as predators: A short journey among behavioural flexibilities, adaptations and feeding habits. *Frontiers in Physiology* 8: 598.
76. Webster, J. (1994). *Animal Welfare: A cool eye towards Eden*. Oxford: Blackwell Science.
77. Webster, J. (2005). *Animal Welfare: Limping towards Eden*. Oxford: Blackwell Science.
78. Webster, J. (2013). *Animal Husbandry Regained: The place of farm animals in sustainable agriculture*. Oxon: Routledge.
79. Webster, J. (2020). *Understanding the Dairy Cow,* 3e. Oxford: Wiley.
80. Whay, H.R., Main, D.C.J., Green, L.E. et al. (2007). Assessment of the behaviour and welfare of laying hens on free-range units. *Veterinary Record* 161: 119–128.
81. Whittingham, L.A. and Dunn, P.O. (2005). Effects of extra-pair and within-pair reproductive success on the opportunity for selection in birds. *Behavioural Ecology* 16:138–144.
82. Wilschko, R. and Wiltschko, W. (2019). Magnetoreception in birds. *Journal of the Royal Society Interface* 16(158): 20190295.
83. Wilson, D. (2015). *The Welfare of Performing Animals; A Historical Perspective*. London: Springer.
84. Wimpenny, J.H., Weir, A.A.S., Clayton, L. et al. (2009). Cognitive procedures associated with sequential tool use in New Caledonian Crows. *PLUSONE* 4(8): e6471.
85. Wittgenstein, L. (1953). *Philosophical Investigations* (English translation). Oxford: Blackwell.

General Reading

Ackerman, J. (2017). *The Genius of Birds*. Penguin Books. –
Benz-Schwarzburg, J. and Knight, A. (2011). Cognitive relations yet moral strangers? *Journal of Animal Ethics* 1: 9–36.
Brabazon, J. (1975). *Albert Schweitzer: A Biography*. Syracuse University Press.
Broom, D.B. and Fraser, A.F. (2007). *Domestic Animal Behaviour and Welfare*. Oxford: CAB International.
Colgan, P. (1989). *Animal Motivation*. London: Chapman and Hall.
Dawkins, M. (1993). *Through our eyes only: the search for animal consciousness*. Oxford: Freeman.
Frith, C. and Frith, U. (2005). Theory of Mind. *Current Biology* 15: 644–646.
Gauthreaux, S.A. (1980). *Animal Migration, Orientation, and Navigation*. Cambridge: Academic Press.
Harrison, R. (2013). *Animal Machines. The New Factory Farming Industry*. Oxford: CAB International.
McGilchrist, I. (2012). *The Master and his Emissary. The Divided Brain and the Making of the Western World*. New Haven: Yale University Press.
McGreevy, P. and McLean, A. (2011). *Equitation Science*. Oxford: Wiley Blackwell.
Singer, P. (1990). *Animal Liberation: A New Ethics for our treatment of Animals*. New York: Avon.
Smith, G. (2020). *Metazoa: Animal Minds and the birth of consciousness*. London: William Collins.
Webster, J. (1994). *Animal Welfare: A cool eye towards Eden*. Oxford: Blackwell Science.
Webster, J. (2005). *Animal Welfare: Limping towards Eden*. Oxford: Blackwell Science.
Webster, J. (2013). *Animal Husbandry Regained: The place of farm animals in sustainable agriculture*. Oxon: Routledge.

Animal Welfare: Understanding Sentient Minds and Why it Matters, First Edition. John Webster.

Index